The LUCAYAN ISLANDS

BOOK 1 of 3

THE LUCAYAN SEA: A Case for Naming the Historic Waters of The Bahamas & Turks and Caicos Islands Series

TELLIS A. BETHEL

Copyrights

ISBN: 9798726574400
Publisher: Inspire Publishing, The Bahamas

www.Inspirepublishing.org

Cover Design: Teri M. Bethel

Disclaimer: The opinions expressed in this manuscript are the opinions of the author and do not represent the opinions or thoughts of the publisher. Because of the internet's dynamic nature, any web addresses or links contained in this book may have changed since publication and may no longer be valid.

Book Cover Image: Photograph taken by an astronaut aboard the International Space Station. The photograph shows the Great Exuma Island cays in The Bahamas with "prominent tidal channels cutting between them. For astronauts, this is one of the most recognizable points on the planet. Image Credit: NASA.

Books written under the series titled *Lucayan Sea: A Case for Naming the Historic Waters of The Bahamas & Turks and Caicos Islands* may be ordered through booksellers and Amazon.com.

Author Contact: Tellis A. Bethel
P. O. Box CB-11990, Nassau, Bahamas
Email: tbethel@tellisbethel.com (www.tellisbethel.com)
Website: www.lucayansea.com

Dedication

To my lovely wife, Teri, and our two wonderful sons, Drue and Tate, with love beyond measure.

Acknowledgments

I'm grateful for the inspiration received from family and friends to promote the idea of giving the historic waters of The Bahamas and Turks and Caicos Islands a name that exemplifies these islands' rich heritage. This series (*The Lucayan Sea: A Case for Naming the Historic Waters of The Bahamas & Turks and Caicos Islands*) and its companion book (How The Bahamas and Turks and Caicos Islands got their Names) is another step forward in this direction.

For this reason, I'm incredibly thankful to Diane Phillips, President of Diane Phillips and Associates, for her enthusiasm and commitment to creating public awareness and support. Di was the first person in the public arena informed of the idea in 2015 and has volunteered her time and expertise in promoting this endeavor ever since. Beyond this, Di and her husband, Larry, are dear friends of my wife, Teri, and me.

It also gives me pleasure to express profound appreciation for Sir Durward Knowles (now deceased) being among the first to publicly endorse the concept of naming our waters the Lucayan Sea. Sir Durward was in his 90s when Di introduced me to him in 2015. He was an avid Bahamian sailor whose superior skills earned him the name "Sea Wolf." Sir Durward and fellow Bahamian teammate Cecil Cooke became the first Bahamians to win an Olympic gold medal in the Star Class Sailing event at the

Tokyo Japan Olympic Games in 1964. I had the privilege of serving as Commodore of the Annual Sir Durward Knowles Festival of Lights Christmas Boat Parade in 2014 and 2015. The event was held in Nassau Harbour in Sir Durward's honor.

I also convey gratitude to Ms. Shonel Ferguson, Member of Parliament, and Chairman of the Clifton Heritage Authority, who welcomed the idea of naming our waters. Ms. Ferguson oversaw the groundbreaking ceremony to construct a Lucayan village at the Clifton Heritage National Park located at the western end of New Providence in 2019. Others who endorsed the concept from inception included: attorney and former President of The Bahamas National Trust, Mr. Pericles Maillis; the President of The Bahamas Historical Society, the Executive Director, Bahamas National Trust, Mr. Eric Carey; the President of The Bahamas Historical Society, Mrs. Andrea Major; the Rev. Canon S. Sebastian Campbell, former Chairman of National Heroes Committee; and Mr. Freddie Munnings Jr., a former member of National Heroes Committee and cultural icon.

I'm also truly grateful for the prayers and support of Apostle Benjamin Smith, Senior Pastor, The Embassy International, and Dr. Deborah Bartlett, President, CEO Network. I'm deeply appreciative of the time taken by Dr. Grace Turner, Senior Archaeologist and Research Officer at the Antiquities, Monuments and Museum Corporation in The Bahamas, and Dr. Christopher Curry, Associate Professor in the History Department at the University of The Bahamas, in providing valuable feedback on the first book in this three-book series (*The Lucayan Islands*).

Indeed, it is an honor to acknowledge four individuals I was privileged to have met during my younger years. Each was a leader in their respective area of expertise. Three of them are now deceased. Their life stories have significantly influenced the writing of my first book, *The Lucayan Sea—Birthplace of the Modern Americas* (no longer in print) in 2015, and now this expanded series.

The first of these individuals is the late Captain Jacques-Yves Cousteau, whom I met during the early 1990s on the island of Eleuthera—the birthplace of the Bahamas. Captain Cousteau was a former French naval officer, co-inventor of the aqualung, and pioneer of Self-contained Underwater Breathing Apparatus (SCUBA) diving in 1943. He was also a forerunner in the underwater exploration and filmmaking industries. In 1985, Captain Cousteau was awarded the U.S. Presidential Medal of Freedom, the United States' highest civilian award for his outstanding contributions to underwater exploration. Captain Cousteau died in 1997.

I was fortunate to have accidentally met Captain Cousteau at the Pine Street and Banks Road intersection near the top of the hill overlooking Governor's Harbour and the quaint community of Cupid's Cay. This sea-world explorer once resided on Windermere Island in Central Eleuthera, a former vacation spot for England's Prince Charles and Lady Diana (now deceased) during the early 1980s. The fascinating conversation we had about his love of scuba diving and his daring adventures into the unknown underwater world gave me an even greater appreciation for the waters of The Bahamas.

I also wish to express my admiration for Mr. George Huntington Hartford II's (now deceased) vision to transform Hog Island, where hogs were raised, into a world-renowned tourist resort. While researching Mr. Hartford's whereabouts, I was surprised to discover that the American businessman, philanthropist, and heir to the Great Atlantic and Pacific Tea (A. & P.) Company resided in The Bahamas. Teri and I enjoyed meeting him at his daughter's home in Lyford Cay on New Providence Island during the early 2000s. He was in his early 90s then, with an effervescent and welcoming personality.

In 1959, Mr. Huntington purchased the 700-acre Hog Island,[1] just a quarter-mile (0.8 km) north of Nassau Harbor, from Swedish Industrialist and multi-millionaire Mr. Axel Wenner-Gren.[2] Previously named Hog Island, Huntington renamed the four-mile

(6.4 km) long island Paradise Island and opened a 52-room luxury hotel named the Ocean Club in 1962. The hotel included a beach, a 12th-century French cloister, gardens, an 18-hole golf course, and a marina with a boat landing area located on New Providence Island's north shore as there was no bridge joining the two islands at the time. American pop star Elvis Presley, the English rock band, the Beatles, and the original James Bond actor, Sean Connery, were among the resorts' guests.

It was also Mr. Hartford's dream to develop the upscale resort as a symbol of peace.[3] His desire reflects one of humanity's highest aspirations, whose origin is told in 17th-century English poet John Milton's (1608–1674) epic poem, Paradise Lost.[4] Milton's work retold the story of how Adam and Eve had stepped out of bounds in the Garden of Eden and were subsequently removed from their earthly paradise. Humankind has been in pursuit of peace ever since, with their ancient trek ending in these Lucayan islands before the dawning of the modern era.

Although Hartford's property would later come under new ownership from as early as 1964, the evolution of his vision of a tropical Paradise has helped transform The Bahamas into a premier tourist destination. Paradise Island is now home to the Atlantis Paradise Island Resort, the region's largest water park, and the world's largest open-air marine habitat.[5] Atlantis is east of Nassau Cruise Port and is visible from cruise ships moored at Port Nassau. Access to Atlantis is by ferry from the Nassau Cruise Port or by Paradise Islands' Sir Sidney Poitier Bridge (named after Bahamian-American Oscar-winning actor). For many visitors, the resort and its amenities have created an unforgettable experience of "Paradise Found,"

The third person I wish to acknowledge is Dr. Myles Munroe, also deceased. Dr. Munroe was a Bahamian international public speaker, author, government consultant, and religious leader. He was a visionary who had traveled the world and inspired hundreds of thousands to maximize their potential in life. Coincidentally, my first visit to the Turks and Caicos Islands was with Dr. Munroe in 1989 during one of his speaking engagements. Dr. Munroe's life and legacy have encouraged Teri and me to pursue our God-given dreams and aspirations for the betterment of others.

I also wish to salute Captain Daniel Zenicazalaya, a Spaniard from Bilbao, the de facto capital of Basque Country in northern Spain. Captain Daniel was the captain of the S/S Emerald Seas (from the 1970s to 1990s). Because of him, I had the opportunity to work aboard this former World War II military troop transport vessel as a young bellboy (bellhop) in 1978. The Emerald Seas was the forerunner of today's mega cruise ships. Unlike Christopher Columbus, this Spaniard (Captain Daniel) transported thousands of cruise ship passengers to The Bahamas to enjoy its sun, sand, and sea. The 600-foot (183 meters) ship was home-ported in Miami, Florida, and operated between Miami, Freeport (Grand Bahama Island), Little Stirrup Cay (renamed CocoCay), and Nassau from the 1970s to 1992.[6]

Admiral Cruises operated the Emerald Seas before Royal Caribbean Cruise Line purchased it in 1988. Little Stirrup Cay is now operated by Royal Caribbean Cruise Line and renamed CocoCay. The tiny cay is located at the northern end of the Berry Island chain in the northern Bahamas. Today, Royal Caribbean Cruise Line is one of the world's largest cruise line operators. Coincidentally, Royal Caribbean's Chief Executive Officer, Michael Bayley, was also employed aboard the Emerald Seas while Captain Daniel was in charge.

Later, I was encouraged by Captain Daniel to pursue a career as a naval officer in the Royal Bahamas Defence Force. Five years after my stint as a bellboy on the Emerald Seas, I recall seeing the Emerald Seas near Little Stirrup Cay (CocoCay) and hailing Captain Daniel over his ship's radio. By this time, I had already completed my initial naval officer training at Britannia Royal Naval College in Devon, England, and was in command of a 60-foot (18 meters) coastal patrol vessel. Captain Daniel still lives in Nassau and was married to a wonderful Bahamian lady, Marlene (nee Thompson), now deceased. The Defence Force's motto is: "Guard Our Heritage," a mantra I have lived by for over 30 years and continue to uphold even after completing my service in the nation's armed service.

I must say that the support received from family, friends, and well-wishers (locally and abroad) during the 18 months of research for this trilogy has deepened my appreciation for the shared heritage of The Bahamas and Turks and Caicos Islands and the potential they possess to impact the world for the better. For this, I am grateful.

Table of Contents

Image Log

Photos

Illustrations

Preface

The world is becoming increasingly rife with civil unrest, disease, wars, natural disasters, and other human and environmental ills. Even here in The Bahamas and Turks and Caicos Islands, Bahamians and Turks and Caicos Islanders are slowly drifting away from rare qualities that made them unique and transformed their islands' rural farming and fishing village communities into world-renowned tourist destinations. Listed high among their priceless attributes is an inherent "people-to-people" friendliness accentuated by a seascape of breath-taking beauty, creating an aura of peace.

Surprisingly, as the world changes, there is one thing that Bahamians and Turks and Caicos Islanders can do that can make their island communities and the world around them a better place. This "one thing" is to give their waters a name that encompasses their islands' rare heritage. The simple act of naming these waters has the potential of developing a "peace culture" within The Bahamas' and Turks and Caicos Islands' communities while promoting sustained economic growth throughout their islands.

The story behind the name this series proposes for these waters could also become a source of inspiration that could change the world for the better. Interestingly, The Bahamas and Turks and Caicos Islands belong to the Lucayan Archipelago, whose waters became the geographic womb from which the Americas' modern nations unfolded after Italian explorer Christopher Columbus arrived off their shores in 1492. As the "birthplace of the Americas," it is believed that the name proposed for these Lucayan waters could also establish them as a "hemispheric" symbol of peace. This positioning could ultimately transform The Bahamas and the Turks and Caicos Islands into the "Peace Capital of the Americas." As such, these islands would act as a "peace preserve" where world bodies and organizations come to resolve critical challenges for the advancement of peace. The idea of a Peace Capital is an integral part of these islands' past, as explained in the third book in this series.

After his first step on the Moon in 1969 in an area where a sea once existed called the Sea of Tranquility (Mare Tranquillitatis),[7] U.S. astronaut Neil Armstrong said, "That's one small step for man, one giant leap for mankind." To give the historic Lucayan waters a name that embraces their heritage and tells their story would be "one small step for Bahamians and Turks and Caicos Islanders, one giant leap for humanity." Amazingly, if this idea is accepted, "it won't cost a dime" to name these waters.

Who is this Series For?

If you're a Bahamian, Turks and Caicos Islander, or one of our residents;

If you're one of our many cruise ship visitors or hotel

Guests;

If you've never visited these islands, but you're an "I love the Bahamas" or "I love the Turks and Caicos Islands" fan;

If you enjoy discovering fresh insights about old places like The Bahamas and Turks and Caicos Islands that have shaped the course of world history;

If you want to be a part of history in the making;

If you're an astronaut, aircraft pilot, a boater, a scuba diver, or anyone who enjoys exploring the great outdoors or simply relaxing;

If you want to do your part to make the world a better place;

then this series is for you.

Author's Note

Humanity's quest for peace has been intrinsically linked to these islands' waters since pre-Columbian times. Throughout history, great stories are told of fascinating achievements that have improved humanity's wellbeing because of this longing. Among these accomplishments were breakthroughs in science, technology, communications, agriculture, medical cures, and much more.

Conversely, humanity's dark side has also been chronicled throughout the ages. Gruesome acts of oppression, injustice, and depravity have produced unimaginable fear, terror, dread, and human demise. Though technology and science have improved exponentially, "humanity's inhumanity to humanity" has also escalated to the point where the potential for self-extermination is now a reality.

A common cause for this great divide between humankind's greatness and depravity is its unbridled pursuit of power and wealth. Nations have often set out to achieve greatness at the expense of freedom (vital for making the world a better place), justice (essential for sustaining life), and peace (necessary for experiencing fulfillment in life). Usually, it is the smaller, more vulnerable nations or people groups that suffer. The misuse of power and wealth has severely diminished prospects for cultivating longevity, prosperity, and security among the nations of today's world. Yet, amid life's triumphs and tragedies, prospects for peace still flicker in the dark.

This series shares much about the devastation unleashed by early European explorers upon indigenous civilizations in the Ancient World. Expansionism, wars, colonialization, and slavery were vehicles for the horrific events that transpired. In bringing balance to perspective, it should be noted that life in the Ancient World was a far cry from being a "utopian" experience among many ethnic groups.

Like Old World Europeans and Africans, the indigenous peoples of the Ancient World, such as the Aztecs, Incas, and Tainos, also engaged in expansionism, wars, and enslavement against neighboring tribes and nations. Even in today's world, conflicts, wars, and slavery still occur among countries and between ethnic groups. These actions should never be condoned. However, care should be taken not to "throw the baby out with the bathwater" (or discard the good that others have accomplished in light of the pain they caused).

Learning from these islands' early experiences during the Americas' founding is critical for preventing the recurrence of past atrocities. For this reason, this trilogy also proposes how Bahamians and Turks and Caicos Islanders can make judicious use of their unique heritage for humanity's betterment.

Terminologies & Pronunciations

Throughout this series, references are made to such terms as the "New World," "the Americas," "the birthplace of the Americas," "the Lucayan Islands," "the Lucayan Archipelago," and "the modern era," and others. These terminologies are explained below to bring clarity to the context in which they are presented.

Pronunciations

Before outlining the terminologies used in this series, there are three keywords that some readers may want to know how to pronounce. These words are Lucayan, Bahamian, and Caicos.

Lucayan is pronounced: **Lu** (as in **Lu**ke)-**kay** (as in **kay**ak)-**an** (as in Puerto Ric**an**).

Bahamian is pronounced: **Ba (**as in **Bah**rain**)-ham** (as in low sounding **h+aim** or **haim**; the **ian** (as in the Japanese **yen**).

Caicos: Cai (as in the letter **K**)-cos (as in **cus**tomer or as in **cos**t**)**

The Ancient World of the Western Hemisphere:

It is widely believed that the continents and islands of the Western Hemisphere were initially settled by Asian peoples and their descendants thousands of years before European arrival. There were no collective names for the Western Hemisphere's continents and archipelagos at the time of the Americas' founding. Although early inhabitants had indigenous names for their ethnic groups, they were often referred to as Amerindians,

Aborigines, Native Americans, Indigenous Peoples, or the First Nations, Inuit and Métis peoples of Canada by Europeans and their descendants. Some of these terminologies are used in this series.

Readers should also note that the name "Americas" never existed until the early 1500s. Consequently, the term "Ancient World of the Western Hemisphere" (or Ancient World in its shortened form) is used in reference to the Western Hemisphere's mainland (the northern and southern continents), islands (circum-Caribbean region and the Lucayan Islands), associated waters, and the indigenous peoples who inhabited these lands before European contact in 1492.

The Lucayan Islands:

"The Lucayan Islands" constitute an archipelago whose northern end is near the southeastern limits of the United States, and its southern boundary is near the northern coasts of the Republic of Haiti and the Dominican Republic on the island of Hispaniola in the northern Caribbean (the Greater Antilles). The Lucayan Archipelago is geographically separated from the United States and the Caribbean.

The Lucayan people are descendants of the Taino people in the northern Caribbean (Greater Antilles), who originally inhabited the Lucayan Archipelago before the Lucayan culture was formed. Today, the archipelago comprises the independent country of the Commonwealth of The Bahamas and the British Overseas Territory of the Turks and Caicos Islands at the island chain's southeastern end.[8]

Whenever the term "Lucayan Islands" or "Lucayan Archipelago" is used in this series, it refers to both the Bahama Islands (The Bahamas) and the Turks and Caicos Islands. When addressing the two territories as politically separated archipelagos, the Bahama Islands is referred to as the Bahama Archipelago, and the Turks and Caicos Islands as the Turks and Caicos Islands Archipelago.

The Modern Era (or Modern History):

"The modern era" is a term that describes the period between 1492 to the present time. This period indicates the beginning of a new era in human history, [9] including the Ancient World of the Western Hemisphere. For Europeans, 1492 was the closing chapter of the Middle Ages (or the medieval period) and the onset of the modern age.[10] The Middle Ages generally refers to the period starting with Rome's fall (476 BC) to the Renaissance Period, which some propose begun during the 13th, 14th, or 15th centuries.[11]

Spain sometimes refers to 1492 as the end of the Middle Ages and the beginning of the modern era.[12] This series identifies the year of Columbus' landfall in the Ancient World (1492) as the closing chapter for the Ancient World of the Western Hemisphere and the beginning of the modern era for the Americas—politically, economically, and culturally.

The "New World" Concept:

The title "New World" was a concept created by Europeans who had a Eurocentric mindset about the world. The Ancient World was "new" to Old World Europeans after Columbus happened upon it. These explorers had no prior knowledge of the Ancient World's existence at the onset of the modern era. However, the Ancient World was not "new" to the Western Hemisphere's indigenous peoples, nor was it "discovered" by Europeans, as indeed it was inhabited by indigenous peoples for thousands of years before European contact.

The Old World of Europe

The term "Old World" generally refers to those continents and peoples on the Atlantic Ocean's eastern side (Europe, Asia, Africa, and Australia).[13] For this book's purposes, the term Old World of Europe is applied to those European countries that colonized the Ancient World after 12 October 1492, resulting in what is now known as the Americas. These countries included Spain, Portugal, England, France, and the (Dutch) Netherlands.

The Americas:

The name Americas is an Old World description of the islands and continents that European explorers encountered in the Ancient World. European cartographers introduced this name at the beginning of a modern era for the Americas during the early 1500s. In this series, the term "Americas" or the "Wider Americas" generally refers to today's North, South, and Central America, and the circum-Caribbean region (including the Lucayan Islands), and associated waters within the Western Hemisphere.

The New World of the Americas:

Although initially "new" to early Europeans after Columbus' first landfall in the Ancient World of the Western Hemisphere, the Ancient World was gradually transformed into a new one politically, economically, culturally, and socially. This transformation began on 12 October 1492, the day Columbus arrived in the Lucayan Islands during his first voyage, beginning the Old World's exploration, exploitation, and colonization of the Ancient World. It was not until the early 1500s that European cartographers named the Ancient World "America."

As a result, this series describes the post-1492 Ancient World as the "New World of the Americas," which consists of the nations that emerged due to the permanent reunion of two main branches of civilization (the Ancient World and the Old World) during the modern era. In the context of this series, the term New World of the Americas does not necessarily apply to those territories inhabited or administered by today's descendants of the indigenous peoples.

Currently, the United Nations (U.N.) acknowledges over 30 sovereign nations in the Western Hemisphere. Territories controlled by indigenous groups within these sovereign nations are not identified as sovereign states by the U.N. Over the years, the Americas has become a melting pot of the Ancient World indigenous peoples and descendants of the Old World peoples (Africans, Europeans, and Asians).

Today, the Americas consists of over 33 modern nations, some of which have allocated lands to descendants of native peoples from the Ancient World to administrate and develop. These territorial allotments include Indian Reservations in the United States, The Toronto Purchase in Canada,[14] and the Kalinago Territory in the Commonwealth of Dominica managed by state-recognized tribes. Nevertheless, much of the indigenous traditions, ways, customs, tribes, and empires were destroyed by the European invasion of the hemisphere, and new nations were born.

The Birthplace of the Americas:

The waters surrounding the Lucayan Archipelago became the gateway for European access to the Ancient World. The events that followed resulted in the Americas' founding and the unfolding of the Americas' modern nations. Therefore, the Lucayan waters surrounding The Bahamas and Turks and Caicos Islands (the Lucayan Archipelago) are described as the "Birthplace of the Americas" (or the "Birthplace of the New World of the Americas") in this series.

The Site Where the Americas were Founded:

"The site where the Americas were founded" refers to the Lucayan Archipelago, where Columbus first arrived and made landfall in the Ancient World in 1492. It was from this site that the modern nations of the Americas unfolded.

Heritage:

The word "heritage" describes the history and geography of an island (or a country) and the culture its people inherited or passed down to successive generations. "History" refers to past events, including political, economic, and social affairs. "Geography" describes the land and sea and an island's natural resources and physical features (or country).

"Culture" is defined as a people's way of life, including customs and traditions. In this series, a national heritage is considered those dimensions of history, geography, and culture beneficial to fulfilling national interests and advancing humanity's longevity, prosperity, and security.

Spaniards and Spanish:

The terms Spaniard or Spanish apply to people, places, or things originating from or related to Spain unless otherwise stated.

Introduction to the Book Series

Despite the horrifying events that culminated in the Americas' founding, today's Bahamians and Turks and Caicos Islanders were left with a rich heritage, along with an untapped potential that could help humanity fulfill its long-standing quest for peace. In taking an unconventional approach towards unveiling this unique heritage, it is believed that the uncapping of this potential can begin with naming these islands' waters.

While the audible appeal and historical relevance of a likely name are important, the selected name should encompass these islands' extraordinary history and geography and the traditionally friendly culture of the people it represents. Consequently, the story behind the selected name ought to make a telling difference in the lives of those to whom it is told.

This series tells the much overlooked yet life-transforming story behind the proposed name for these islands' waters. This fascinating story also highlights the incredible attributes of The Bahamas' and Turks and Caicos Islands' "hemispheric" heritage. Beyond giving birth to the Americas, this profound heritage is latent with qualities that foster and promote peace, unfolding what can be termed "the Lucayan Sea Story."

Today, The Bahamas and Turks and Caicos Islands are hailed as premier tourist destinations. They are described as "jewels" in the Caribbean region because of their friendly people and beautiful waters within a peaceful tropical environment. Nevertheless, beneath these islands' tranquil setting of "tropical sun, sand, and sea" are telltale signs of past human atrocities that gave rise to the Americas' modern nations.

The awe-inspiring waters that encompass the Lucayan Archipelago bore witness to the brunt of these tragic events during the Americas' founding. Today, these waters stand as a reminder for those who behold them to do their part to make the world a more peaceful place.

The Lucayan Sea: A Case for Naming the Historic Waters of The Bahamas & Turks and Caicos Islands series tells the story of the Americas' founding, with a fresh look at the past. Learning from the past often shines new light on current experiences that can uncover unprecedented ways of making the best use of one's national heritage.

This series takes an unconventional look at remarkable facets of these islands' history, geography, and culture during the Americas' early founding. These rare aspects of The Bahamas and Turks and Caicos Islands heritage are woven together to create a framework for promoting peace throughout these islands, the Americas, and the world.

In telling the Lucayan Sea story, insights provided are primarily derived from events resulting from Columbus' arrival in the Lucayan Islands and Spain's colonization of the Americas. The unique insights discovered within these islands' heritage that can promote peace are presented as proposals throughout this series. Propositions include giving the Lucayan waters a suitable name, promoting these waters as the birthplace of the Americas and a symbol of peace, branding these waters for economic diversification, and establishing the Lucayan Islands as the Peace Capital of the Americas.

Admittedly, this writer considers himself a "history enthusiast" with a keen interest in The Bahamas' and Turks and Caicos Islands' world-changing heritage, but not a historian.

Nevertheless, this author has intimate knowledge of the Bahama Islands and their Lucayan waters (as a naval officer) with over 30 years in a military organization whose motto is: Guard Our Heritage. For this writer, being the "birthplace of the Americas," with an "inherent responsibility" to make the world a more peaceful place is a heritage worth protecting.

The first book in this series, *The Lucayan Islands*, shares insights into The Bahamas' and Turks and Caicos Islands' heritage from the initial populating of Ancient World of the Western Hemisphere and the settling of the Caribbean and Lucayan Islands to Columbus' "discovery of the Americas."

The second book, *The Lucayan Sea—Birthplace of the Americas*, gives an overview of how the Spanish conquered the Ancient World and transformed it into the New World of the Americas. It also reveals how the Lucayan waters became the birthplace of the Americas, with the Lucayans becoming the first to suffer genocide at the beginning of a new era in human history.

Additionally, this book identifies the main waterways adjacent to the Lucayan Islands that shaped the Americas' founding and how these waters got their names. Further, it proposes a name for The Bahamas' and the Turks and Caicos Islands' historic waters, along with reasons for selecting it.

The third book in this series, *The Bahamas and Turks and Caicos Islands—Peace Capital of the Americas*, suggests a leadership role Bahamians and Turks and Caicos Islanders were destined to play as facilitators of peace. It further recommends the branding of these islands' waters for economic prosperity and community development. This book also defines what "peace" means in the context of these islands' heritage.

Additionally, this third book proposes a maritime alliance between The Bahamas and the Turks and Caicos Islands under the proposed name for the Lucayan waters. It highlights this alliance as a platform for strengthening economic, cultural, and security ties between The Bahamas and Turks and Caicos

Islands. *The Bahamas and Turks and Caicos Islands—Peace Capital of the Americas* concludes with a conceptual framework for transforming these islands into the "Peace Capital of the Americas."

This series can also supplement other interesting books about The Bahamas and Turks and Caicos Islands written by seasoned Bahamian, Turks and Caicos Islander, and foreign authors. This writer is grateful for these authors' published works cited throughout this series for those wishing to engage in further reading.

The books in this series were also written to accommodate a cross-section of readers throughout The Bahamas and Turks and Caicos Islands, and beyond. Therefore, scientific jargon is kept at a minimum wherever possible. Readers should note that this series elaborates on contents drawn from books previously written by this author. These books include *The Lucayan Sea: Birthplace of the Modern Americas* (no longer in print) and *The Lucayan Story* (for high school and college students).

Complementing this trilogy is the mini-book, *How the Bahamas & Turks and Caicos Islands got their Names.* This companion book draws insights from the Lucayan Islands' topography, geography, the Spanish mother-tongue, Spanish-American culture, and the early Spanish explorers' experiences to provide possible answers to how the two countries were named.

Therefore, the perspectives and suggestions presented in this series create opportunities for further study, constructive debate, and insightful dialogue for the best use of these islands' heritage.

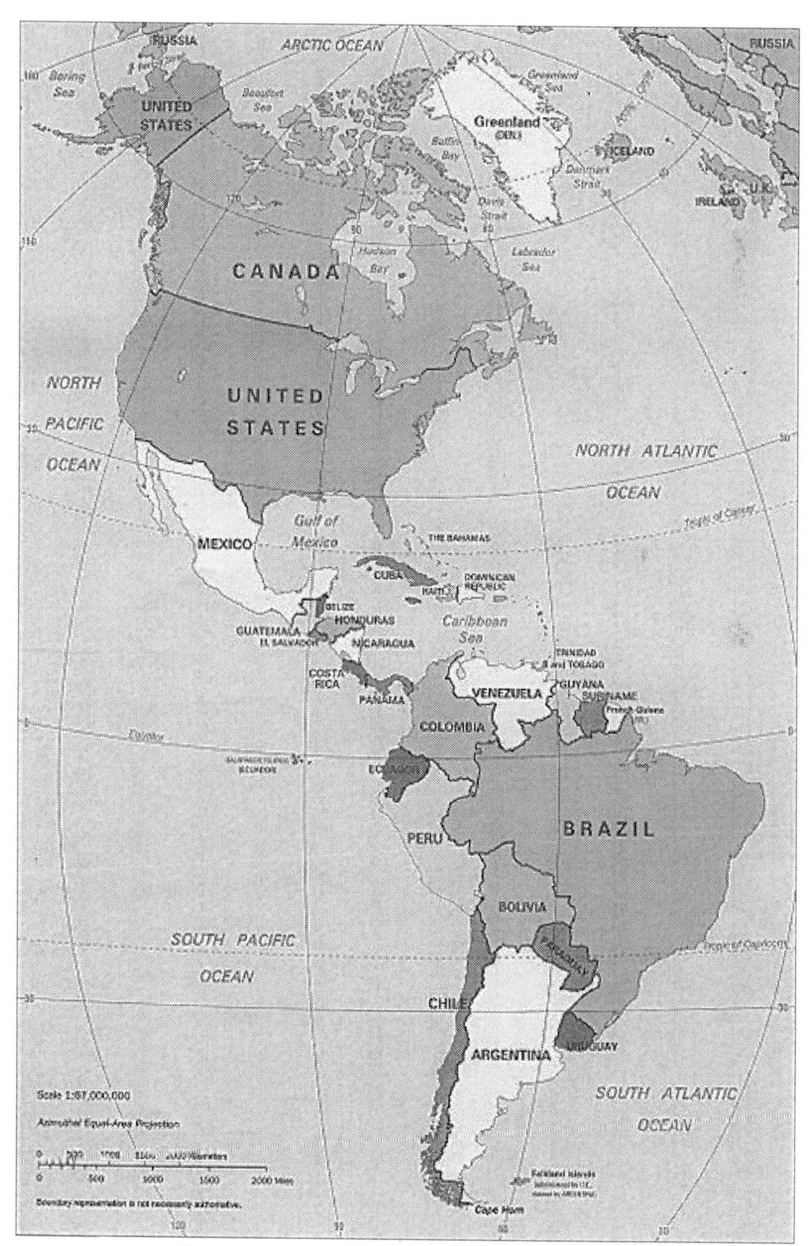

Illustration 1: Map of the Americas (Source: Wikipedia)

Introduction to Book I: The Lucayan Islands

The Lucayan Islands is the first book in this series. This book's name is derived from the Spanish form of the indigenous name for the Lucayan Islands, Las Islas de Los Lucayans. It lays a foundation for Bahamians and Turks and Caicos Islanders to tell their islands' story, revealing unique insights for humanity's betterment through the naming of their waters. As the first in this series, *The Lucayan Islands* takes readers on a voyage of discovery, focusing on the pre-Columbian and early colonial eras of The Bahamas and Turks and Caicos Islands.

As part of the story behind the proposed name for these islands' waters, readers are introduced to the Lucayan Islands' geological and geographical settings, key aspects of government, population, and culture, and the more popular floras and faunas found on or around The Bahamas and Turks and Caicos Islands.

During their voyage through these islands, readers will discover how English colonists from Bermuda resettled these islands, and how the fledgling colonies became The Bahamas and the Turks and Caicos Islands during the early British colonial period (1648 – 1976). The overview reveals how The Bahamas and Turks and Caicos Islands had progressed after being abandoned for more than a century.

Readers are then taken on a journey from pre-historic to the Spanish colonial (1493 – 1898) eras in the Americas. Along the way, they explore how the Ancient World of the Western Hemisphere (now North and South America) was populated, how the Lucayans settled the Lucayan Islands, and the impact the neighboring indigenous groups, the Arawaks, Tainos, and Caribs, had on these islands and the region.

This first book also shares insights into reports of European, African, and Chinese explorers who may have arrived in the Ancient World before Christopher Columbus. The *Lucayan Islands* ends with Columbus' early life, his accidental landfall in the Ancient World, and exploration of the Lucayan Islands before finding gold in Hispaniola, and completing his first of four transatlantic voyages to the Americas.

The Lucayan Islands has some 177 pages of insightful information with over 35 pages of citations and bibliography, which reveal how:

- the Lucayan Islands' original inhabitants represented the end of the trail for the ancient migration out of Asia into the Ancient World;
- the Caribbean people were mistakenly named;
- the Lucayans were the first to influence western culture;
- the Lucayans were the first to be forcibly removed from their homeland following Columbus' first landfall in the Ancient World;
- the special gift Columbus received from a Cacique (local chief) on Hispaniola (today's Haiti and the Dominican Republic) led to the colonization of the Ancient World;
- Spain may not have been the first European country to claim territorial jurisdiction over the Lucayan Islands;
- The "discovery of the tiny Lucayan Islands resulted in global maritime boundaries being divided on two occasions by two superpowers.

To you, our readers, we say: Welcome to *The Lucayan Islands (or Las Islas de Los Lucayos, in Spanish),* the first book in our trilogy titled *The Lucayan Sea: A Case for Naming the Historic Waters of The Bahamas & Turks and Caicos Islands.*

Photo 2: Map of the Caribbean Region (Source: Wikimedia)

1

A Shared Heritage—
Bahamas and Turks & Caicos Islands

The Bahamas' and Turks and Caicos Islands' shared heritage comprises the site where the Old World of Europe began transforming the Ancient World of the Western Hemisphere into the New World of the Americas. This shared heritage has existed since these islands' were geologically formed (as an archipelago) millions of years ago. These islands' heritage continued with their initial settling by the indigenous Taínos, the depopulation of these islands' original inhabitants (the Lucayans), the resettling by British colonists from Bermuda, and the importation of West Africans as slaves during the Transatlantic Slave Trade.

Even after The Bahamas' independence in 1973 and the Turks and Caicos Islands' return to British Crown Colony status that year, the two countries have maintained close familial, cultural, economic, and security ties. Today, The Bahamas and Turks and Caicos Islands make up the Lucayan Archipelago, with over 700 islands on the mid-North Atlantic Ocean's western perimeter. Both countries are located within the Caribbean region, just north of Hispaniola (the Dominican Republic and the Republic of Haiti) and the Republic of Cuba.[15]

The two countries are now part of the British West Indies, a geopolitical area in the Caribbean region consisting of former

British colonies and current Overseas Territories. Additionally, they are members of the British Commonwealth of Nations, with the British Monarch as the ceremonial head of state. The British Commonwealth is a body of sovereign states that were once dependencies of Britain; these countries now work together for the common good of member states.[16]

Although both countries have been politically separated for over 40 years, references to The Bahamas and Turks and Caicos Islands are often used interchangeably. This intermixing is primarily due to similarities in the history, geography, and culture of these islands. For example, both countries are sometimes called the Bahama Archipelago, Lucayan Archipelago, or the Lucayan Islands.

Knowing which country is being addressed can be confusing at times. For this book's purposes and the sake of clarity, the two countries are individually referred to by their proper or shortened names, The Bahamas or the Turks and Caicos Islands. Additionally, they are called the Lucayan Islands, the Lucayan Archipelago, or the Lucayan chain (of islands) whenever referring to both countries.

The Bahamas' and Turks and Caicos Islands' heritage was further entrenched after many Turks and Caicos Islanders migrated to The Bahamas for employment during the 1950s and 1960s.[17] Many of these "Belongers" have Bahamian spouses and children and have made significant contributions to The Bahamas' national success. Today, their descendants continue to do the same, enriching the common bond between the two countries.

For this series, readers should note that Bahamians and Turks and Caicos Islanders are regarded as one people, considering their shared heritage and the period this series generally covers (from the pre-Columbian era to political separation in 1973).

Historical Highlight

Astronauts Land on "the Sea of Tranquility"

Humankind's deep-rooted desire for peace was considerably lifted when men first landed on the Moon some 234,474 miles (377,349 km) away from Planet Earth. The pioneering feat was accomplished after Apollo 11 was launched from Florida on an eight-day mission in 1969. The small spacecraft made landfall on the Moon on 20 July 1969 with Commander Neil Armstrong and module pilot Buzz Aldrin aboard.

The spacecraft touched down in an area on the Moon called the Sea of Tranquility (once thought to be an ocean known as Mare Tranquillitatis).[18] The astronauts had no encounters with alien beings during their short voyage and safely returned to Earth. However, before departing, the American astronauts left a plaque behind with the inscription: "Here men from the planet Earth first set foot upon the Moon, July 1969 A.D. We came in peace for all mankind."[19]

This historic event occurred almost 500 years after European explorers embarked on a 69-day maritime expedition from Europe in 1492. Their voyage took them across the Atlantic Ocean from Spain into an ancient world new to them. The Spanish-sponsored explorers made landfall within a group of islands approximately 50 miles (87 km) off the East Florida coast—not far from where Apollo 11 was launched. Here, among the Lucayan Islands, the Spanish were the first Europeans to be warmly welcomed to the Ancient World of the Western Hemisphere by the indigenous Lucayan people.

Photo 3: Astronaut Buzz Aldrin salutes the U.S. flag on Mare Tranquillitatis (Sea of Tranquility) during Apollo 11 voyage to the Moon in 1969. (Source: Wikimedia)

Like the Apollo astronauts who landed on the Sea of Tranquility, the early Europeans who arrived in this ancient world on 12 October 1492 "discovered'" a peaceful sea. They also encountered the friendly Lucayan people. However, unlike the American astronauts, the European navigators did not leave a plaque stating that their mission was one of peace. What they did bring were plagues, enslavement, colonization, wars, and genocide. These adventurers turned the Sea of Tranquility they encountered in the Ancient World into a Sea of Turmoil.

Today, humanity is now exploring the far reaches of the Earth's solar system in its attempt to colonize Mars. As they do, these earthlings leave behind a human race that is currently walking a tightrope with the hope of peace at one end and the fear of self-destruction at the other. In between are the winds of uncertainty. Meanwhile, on the home front, it is believed that the people of The Bahamas and Turks and Caicos Islands have an inherent capability and the capacity to help keep the hope of peace alive.

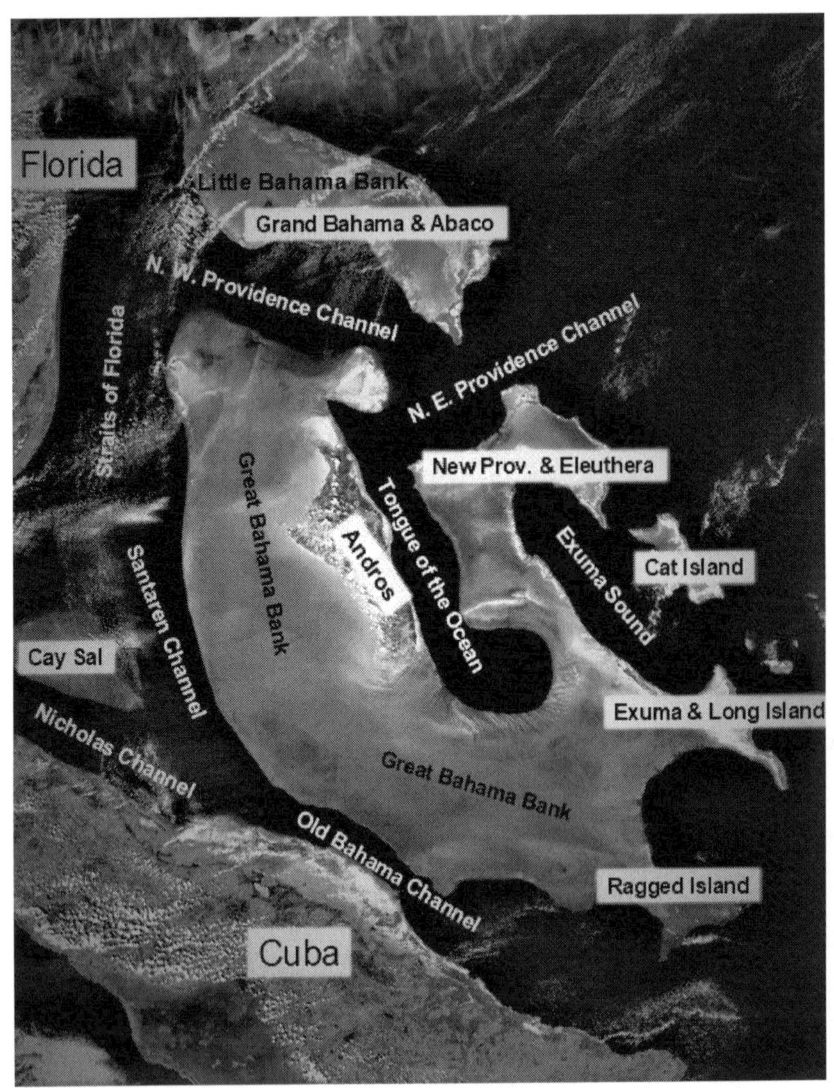

Photo 4: The Tongue of the Ocean and main channels around the Bahama Islands (Source: NASA. Names of islands, channels, banks, and sound were inserted).

2

A Story to Tell

Though barely visible on a large map, The Bahamas and Turks and Caicos Islands have seen more than their fair share of turbulence over the centuries. Yet, these tiny islands' waters have been at the epicenter of changes within the Americas that transformed the world. Events such as the end of the ancient migration's trail out of Asia, the first landfall of European explorers in the Ancient World of the Western Hemisphere, and the initiation of the transformation of the Ancient World into the New World all took place in these islands.

Yet, there is much more to The Bahamas' and Turks and Caicos Islands' story. The name proposed for their waters would help tell this extraordinary story. Left within the wake of the tragic episodes that originated in these islands' waters are clues for fulfilling a call to make the world a better place. Unfortunately, these extraordinary insights remain hidden to many as it appears to some that these islands have little else to offer other than "sun, sand, and sea."

Nineteen years after Columbus' landfall in the Ancient World, a Portuguese historian employed by the Spanish Crown, Peter Martyr Henghera (also spelled D'Anghera), described the Lucayan Islands as "useless" in his book *De Orbe Novo (Of the New World)* written in 1511.[20] Italian explorer Christopher Columbus conveyed this sense of *uselessness* after exploring the Lucayan Islands.

Columbus spent two weeks exploring the Lucayan chain's central and southeastern islands after making his first landfall in the Ancient World in 1492. On completing an unsuccessful search for the Indies' gold and spices, he left, never to return, although he made three more voyages to the Americas.

Unbeknownst to Columbus and Henghera at the time, there is considerably "more to these islands than meets the eyes." Ironically, these hopeless islands became the tiny seed from which the great nations of the Americas grew. Today, the Lucayan Islands possess a rare birthright with massive potential for humanity's benefit if realized.

Upon making landfall on 12 October 1492, Columbus claimed these islands on behalf of the Spanish Crown. The Spanish neglected to colonize them due to their presumed barrenness. There was no gold or silver on these islands, no spice, no pearls, nor silk to be found. The islands were low-lying, with infertile soil unsuited for large-scale farming.

After Columbus' landfall, the Spanish periodically visited the islands to conduct slave raids among the Lucayan people until their population was entirely depleted by the early 16th century. With most Lucayans captured and the few remaining left to die, the Lucayan islands were abandoned for over a hundred years before being resettled by British colonists from Bermuda.

Beyond the slave-raids, Spanish sailors revisited the islands to replenish their water supplies and search for the fabled Fountain of Youth during the 1500s and 1600s. Later, Spanish and French forces repeatedly ransacked and burned the main towns on several islands to the ground. These attacks were retaliation for the island's English settlers' collaboration with notorious pirates who looted Spanish treasure ships and ambushed Spanish ports during the Golden Age of Piracy around the 1600s and 1700s.

In 2011, the *Proceedings of the Fourteenth Symposium on the Natural History of the Bahamas* on San Salvador Island highlighted the fact that The Bahamas is often seen in the light of a single historical event, which is the meeting of two significant branches of civilization—the Old World with the Ancient World.[21] These islands are also typically glamorized as an idyllic paradise or home to the once-notorious Pirates' Republic apart from Columbus' landfall.

Even the archipelago's geography is often mistakenly lumped together with the Caribbean Islands. The Bahamas and Turks and Caicos Islands are erroneously described as being among the Caribbean Islands and their waters as the Caribbean Sea. Moreover, these islands' pivotal role in the Americas' founding and the profound significance their unique heritage was meant to have on humanity's quest for peace is little known and seldom shared or discussed in the myriad of publications and documentaries concerning the early Americas.

Consequently, these islands' unmatched heritage and their potential to change the world for the better remain hidden and are being buried under waves of external influences by countries that rightly promote their heritages. Bahamians and Turks and Caicos Islanders need to rediscover their heritage and tell their story.

Theirs is a story of humanity's quest for peace that gave birth to the Americas, leaving Bahamians and Turks and Caicos Islanders with an inherent responsibility to become facilitators of peace in today's world.

Giving their waters an authentic name that incorporates their heritage is an effective way for Bahamians and Turks and Caicos Islanders to tell their story for the advancement of peace. It is believed that providing these waters such a name would also create an array of unique opportunities that can benefit Bahamians and Turks and Caicos Islanders economically and culturally.

Giving their waters an authentic name that incorporates their heritage is an effective way for Bahamians and Turks and Caicos

Islanders to tell their story for the advancement of peace. It is believed that providing these waters, such a name would also

Photo 5: A view of Atlantis Hotel on Paradise Island from the Royal Caribbean Cruises' Symphony of the Seas (the world's largest cruise ship) moored at the Nassau cruise port in The Bahamas (Source: Tellis Bethel)

create an array of unique opportunities that can benefit Bahamians and Turks and Caicos Islanders economically and culturally.

Coincidentally, the Lucayan chain of islands appears to have been aptly furnished by nature to tell these islands' captivating story. Amid this tropical chain of islands near the North Atlantic Ocean's western border is the Tongue of the Ocean. This underwater trench is over a mile (1.6 km) deep; its depth is comparable to Arizona's Grand Canyon.[22]

The Tongue of the Ocean is surrounded by shallow waters described by astronauts as "the most beautiful place" from space.[23] Its deep gorge with a tongue-like shape (from which its name is derived) cuts into one of the largest banks in the region (the Great Bahama Bank). The Tongue of the Ocean is a marine channel located between the largest Bahamian Island,[24] Andros Island, and New Providence Island, where The Bahamas' capital city, Nassau, is located. (See Photo 5, page 3).

Reaching over 12,000 feet (3,700 m) in-depth, the Tongue of the Ocean runs almost 140 miles (225 km) southward along Andros Island's east coast. The marine trench averages 25 miles (40 km) in width and is home to the Atlantic Underwater Testing and Evaluation Center (AUTEC) for U.S. naval undersea simulated warfare. AUTEC's main base is located in Central Andros.[25]

Although these tranquil waters belie a turbulent past, they speak of a purpose for those who encounter them, and their story reveals a life-preserving identity for their islands' inheritors. Having borne witness to the Lucayan demise, it seems only fitting to dedicate these pristine Lucayan waters to the Lucayans' memory by giving them a name that encompasses these islands' and their inhabitants' history, geography, and culture. This dedication would also serve as a tribute to other indigenous peoples, the enslaved West Africans, and those who suffered or died under the yoke of slavery, establishing these waters as a symbol of peace.

With over seven million annual visitors arriving on their shores, Bahamians and Turks and Caicos Islanders are blessed with the perfect opportunity to tell their story and influence the world for the better. Today, the Tongue of the Ocean stands as a symbolic voice of peace. Amid today's calamities, it appears that this voice now beckons Bahamians and Turks and Caicos Islanders to embrace their heritage, identity, and calling as they tell their story.

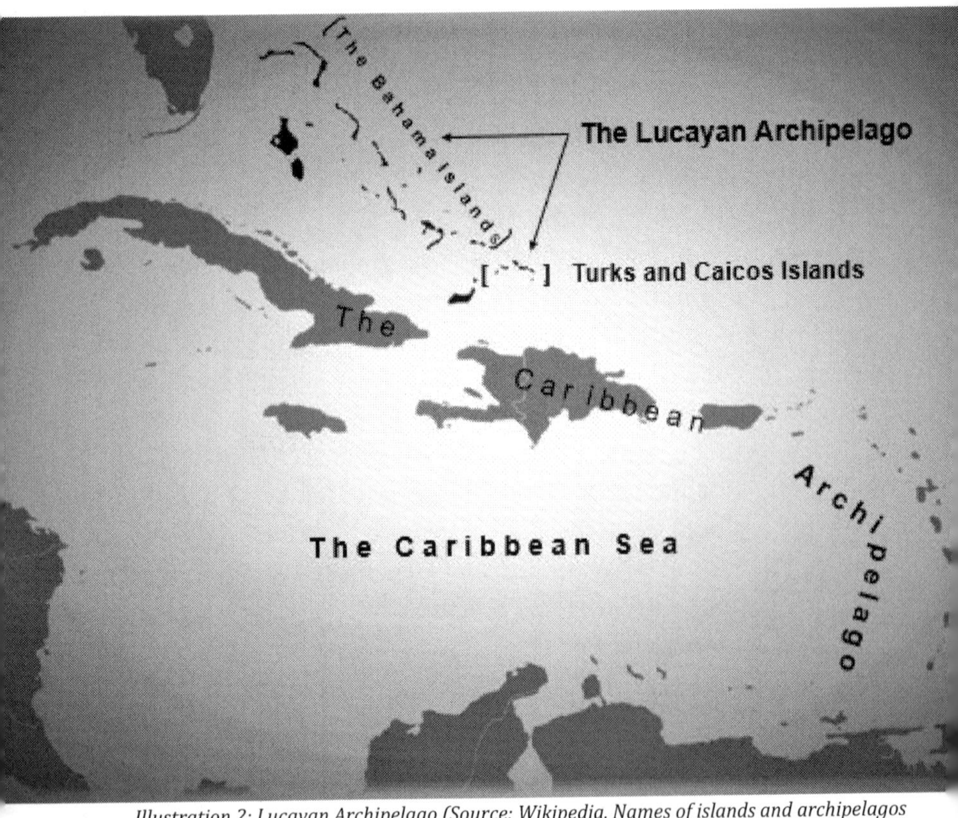

Illustration 2: Lucayan Archipelago (Source: Wikipedia. Names of islands and archipelagos were inserted)

3

Islands of the Lucayans

The word "archipelago" is derived from two Greek words, "arkh" meaning "first" and "pelagos" meaning "sea."[26] Consistent with these definitions are the waters encompassing these islands, as they were the "first sea" sighted by Christopher Columbus. Of equal note, these Lucayan waters are the geographical womb that gave birth to the Americas.

Many people living throughout the Americas and millions who visit these shores are unaware of these islands' geographical location or deep-rooted, historical connections with the Americas' modern nations. In providing this information, this series tells the full story behind the name being suggested for the Lucayan waters and reveals these islands' role in the Americas' founding.

Geographically, the Lucayan Archipelago is located on a northwest-to-southeast axis. This axis extends from approximately 50 miles (80 km) off the southeast coast of the United States of America to within 90 miles (170 km) of Haiti on the island of Hispaniola (also the island of the Dominican Republic) in the northern Caribbean (the Greater Antilles).

The archipelago stretches approximately 600 miles (965 km) from Abaco Island in the northern Bahamas to Inagua Island in the southern Bahamas. Inagua is a Lucayan name, and the island is the second largest in the chain. As The Bahamas' closest point to the Turks and Caicos Islands, Little Inagua is approximately 40 miles (64 km) southwest of West Caicos in the Caicos group of islands. The Lucayan Islands extend about 370 miles (600 km) at their widest points from Cay Sal (the western-most landmass) and San Salvador (the eastern-most island), with both places in the Bahama chain. The Turks and Caicos Islands are about 37 miles (60 km) wide (West Caicos to Grand Turk) and 50 miles (80 km) long from North Caicos to Seal Cays.

Today, The Bahamas comprises all the northern and central islands and most islands in the southern part of the Lucayan chain. Approximately ninety percent of the island chain is Bahamian territory. The Turks and Caicos Islands are a smaller group of islands at the chain's southeastern end. The combined land area of the chain is 5,724 square miles (14,825 sq. km).

The Tropic of Cancer—a circle of latitude at the northern end of the tropics—passes through a beach on Little Exuma Island, at the southeastern end of Great Exuma Island. The beach is known as the Tropic of Cancer Beach and also carries the name Pelican Beach.[27] Pelican Beach is considered one of the prettiest in the Exumas. The Exuma chain consists of over 360 cays and islets famed for having some of the most beautiful beaches in the Lucayan chain, including the famous Pig Beach.

Early Spanish explorers saw the Lucayan Islands as belonging to the same geographical group of islands (or archipelago) and subsequently referred to them as "Lucayos"[28] or the "Lucayas." Las Islas de Los Lucayos[29] was perhaps the most appropriate name the Spanish explorers called the archipelago. The Spanish name was drawn from the indigenous name Lukku Cairi, which the Lucayans called themselves. This indigenous name comes from Taino words, meaning Island People. "Lu," meaning "tribe" or "people," and "cairi," meaning "island." Therefore, Lukku Cairi is believed to mean "Island People."[30]

Lucayos is the Latinized form of the indigenous name. Las Islas de Los Lucayos. It means the *Islands of the Lucayans* (or *the Lucayan Islands*) and was used from the early 1500s.[31] The

French, who explored the region after the Spanish, called the islands "Les Lucayes."[32] European versions of Lukku Cairi (in Spanish: Lucayos or French: Lucayes) remained on nautical charts for over 300 years until the early 1900s.[33]

Today, scholars refer to the island chain comprising The Bahamas and Turks and Caicos islands as the Lucayan Archipelago[34] or the Lucayan Islands,[35] thereby maintaining their historical and etymological roots. These linkages with the past played a key role in recommending a name for these waters that embodies their islands' Taino, Lucayan, Spanish, and English heritage. Details on this historical linkage are explained in greater detail in the second book in this series, The Lucayan Sea—Birthplace of the Americas.

The Europeans called the much larger archipelago south of the Lucayan Islands the Caribbean Islands, highlighting their geographical separation. The Old Bahama Channel at the southern end of the Lucayan Archipelago separates the Lucayan Islands from the northern Caribbean Islands (Cuba and Hispaniola). This channel runs along Cuba's north coast and the northern entrance of the Windward Passage between Cuba and Haiti. Today, the Old Bahama Channel is a primary passage for cruise ships and commercial vessels traveling to and from the American continents (see Photo 4, page 6).

On the western side of the Lucayan Islands are the Florida Straits, once identified as the "Channel of Bahama" on the 1775 map titled *The coast of West Florida and Louisiana. The Peninsula and Gulf of Florida or Channel of Bahama with the Bahama Islands.*[36] The Lucayan Islands are fully exposed to the Atlantic Ocean on their eastern and northern borders.

These islands were formed by plate tectonics (or movements of the Earth's surface) when a single landmass called Pangaea over two million years ago. Geologists believe that the Pangaea's splitting might have taken place around The Bahamas, thereby creating a deep water basin that became known as the Atlantic

Ocean. The continental separation also created the Great Bahama Canyon (also known as the Northeast and Northwest Providence Channels), with canyon walls about three miles (4.8 km) high, believed to be the largest underwater canyon of its kind in the world.[37]

Accumulation of debris (corals, shells, and other deposits) over continental shelves created a plateau equal in size to Texas, known as the Bahama platform.[38] (This platform includes the Turks and Caicos Islands). As the space filled with water, coral and algae began to grow. The decaying corals and shells slowly accumulated over time and eventually developed into the Lucayan Islands.[39] The Bahama platform (including the Turks and Caicos Islands) has maintained its geographical location ever since and is "unrelated to, and geologically quite different from the rest of the West Indies" [40] (that is, other islands in the Caribbean region).

It is believed that the Pleistocene glacial periods resulted in the lowering of seawater levels, which exposed the surfaces of the large continental shelves of these islands. Accumulation of wind-blown deposits later added to the banks' surfaces between glacial periods, creating today's islands. Much of their floras and faunas originated from North America and the Greater Antilles.[41]

The islands, rocks, and cays are the exposed surfaces of large submerged platforms (the Great and Little Bahama Bank) in the central and northern Bahamas. Several smaller isolated carbonate platforms exist in the western and southeastern Bahamas and the Turks and Caicos Islands, such as the Cay Sal Bank in The Bahamas and the Caicos Bank in the Turks and Caicos Islands. Unlike the larger Caribbean Islands that were formed by active volcanoes,[42] the Lucayan Islands are made chiefly of fossil coral and porous (oolitic) limestone surfaces[43] and are north of the Caribbean (tectonic) Plate underlying Central America and the Caribbean Sea.

The northernmost island in the Lucayan chain is Abaco in The Bahamas, and the southernmost island is Salt Cay in the Turks and Caicos Islands, north of Haiti. The nearest islands to the east coast of Florida are Grand Bahama and Bimini in the northern Bahamas, approximately 50 miles (80 km) to the east of Miami and Palm Beach, Florida.

There are isolated submarine platforms that extend farther southeast of Grand Turk in the Turks and Caicos Islands. These platforms include the submerged banks of the Mouchoir, Silver, and Navidad Banks. (The United Kingdom and the Dominican Republic are still negotiating territorial claims regarding the Mouchoir Bank).[44] The Silver and Navidad Banks are under the territorial jurisdiction of the Dominican Republic. The Silver Bank is renowned for hosting one of the world's highest seasonal populations of Humpback whales every winter.[45]

Although the Lucayan Islands have no mountains, the freshwater tidal stream[46] near the western side of Central Andros in The Bahamas is called the Goose River.[47] The source of its freshwater is underground. Andros Island has the archipelago's largest freshwater lens found less than five feet (1.5 meters) beneath the ground.[48] There are no mountains in the island chain. The Lucayan Islands are primarily flat with rolling hills, mainly in the chain's central and southern parts. The highest land elevation in the Lucayan chain is Mount Alvernia, also known as Como Hill, on Cat Island in the central Bahamas. It stands 206 feet (62.8 meters) high.[49]

The Turks and Caicos' highest point is 161 feet (49 meters) in the Blue Hills area on Providenciales Island.[50] Hills in the Lucayan Islands pale in height compared to the 10,164-feet (3,098 meters) tall Pico Duarte Mountain in the Dominican Republic (on Hispaniola)[51] or the 6,476 feet (1,974 meters) tall Pico Turquino Mountain in Cuba.[52] Pico Duarte Mountain is the tallest among the Caribbean Islands.

Combined, the Lucayan Archipelago has a linear coastline that is 2,400 miles (1,491 km) long. The Bahamas has 2,200 miles (3,542 km)[53] of coastline, and the Turks and Caicos Islands' coastline is 242 miles (389 km) long.[54] The combined coastlines are about a hundred miles more than Cuba's, with the longest linear coastline of 2,316 miles (3,735 km) among the Caribbean Islands.[55]

The Bahamas' national waters cover approximately 260,000 square miles (673,396 sq. km) within its exclusive economic zone. Its internal waters consist of large banks, numerous reefs, major shipping lanes, and pelagic fisheries.[56] In 1867, Governor Rawson W. Rawson, after whom the town square in Nassau (Rawson Square) is named, reported the first official numbering of the Lucayan Islands, rocks, and cays. He wrote to British authorities that there were 29 islands, 661 cays, and 2,387 rocks.[57]

The Bahamas occupies roughly 90 percent of the Lucayan chain's landmass with 5,340 square miles (13,830 sq. km) of land. At low tide, the Turks and Caicos Islands' combined landmass is about 366 square miles (948 sq. km).[58]

Andros covers about 3,700 sq. miles (5,900 sq. km) of land and is the largest island within the Lucayan chain, constituting about 65 percent of the total land area. This island is the fifth-largest in the Caribbean region and is divided into three major parts—North, Central, and South Andros. Middle Caicos is the largest of the Turks and Caicos Islands and is roughly 56 sq. miles (144 sq. km).

There are 14 major marine banks in The Bahamas and two marine banks in the Turks and Caicos Islands.[59] Two of the banks, the Great and Little Bahama Banks, are among the largest in the Caribbean region and are located in The Bahamas. The Great Bahama Bank covers the central and southern Bahamas and is the largest of these banks.

It is approximately 37,000 square miles (96,000 sq. km) in size[60] and is about 450 miles (724 km) in length from the central to the southern Bahamas. The Great Bahama Bank is about 170 miles (273 km) at its widest point, from the Florida Straits (near its western border) to the Ragged Island chain on its eastern border.[61] Islands that border this bank include Andros, the Bimini chain, Berry Island chain, New Providence, the Great Exuma Island chain, Long Island, Ragged Island chain, and Eleuthera. Cat Island, Crooked Island, Acklins Island, Inagua Island and Mayaguana Island are exposed surfaces of isolated platforms in the Bahama chain.

This massive bank averages 18 feet (5 meters) to 30 feet (9 meters) in depth and produces an abundance of aragonite mud.[62] The bank also contains one of the largest and purest deposits of oolitic aragonite sand in the region. Private companies export the sand to the U.S., where it is processed, and The Bahamas Government is paid royalties.

Aragonite is used for manufacturing glass, cement, pharmaceutical supplies, and other goods.[63] It is made of calcium carbonate and can be found in abundance on the western side of the 140-mile (225 km) long Andros Island. The sand is named after a town in Spain (Molina de Aragón) where the "rock type was identified" in the 1700s.[64]

There are two main channels in the northern Bahamas—the Northeast and Northwest Providence Channel. The Bahama group of islands is bounded in the south by the Old Bahama Channel. These channels are navigated by merchant vessels and cruise ships transiting the northern Bahamas. At the western entrance of the Old Bahama Channel is the Santaren Channel, which runs along the southwest edge of the Great Bahama Bank and the Nicholas Channel. The Nicholas Channel lies between Cuba's northwestern coast and The Bahamas' western-most islands on the Cay Sal Bank. The Cay Sal Bank is approximately 65 miles (104 km) southeast of Key West, Florida. (See Photo 4, page 6).

The Old Bahama Channel was a major thoroughfare for Spanish fleets entering Havana, Cuba, en route to the American mainland and treasure ships departing Cuba on their return voyages to Spain. The Old Bahama Channel generally runs east-west and is about 434 miles (700 km) long. The channel's width ranges from approximately 17 miles (27 km) to 90 miles (144 km) between Cuba's northern coast and the Great Bahama Bank's southern boundary in The Bahamas. Today, merchant and cruise vessels use the channel to transit The Bahamas to or from Europe, Africa, the Caribbean, or the Americas.

On the western border of the Lucayan chain is the Florida Straits (or the Straits of Florida), a deep water channel that forms the Gulf of Mexico's northern entrance. Mariners often refer to water in this channel as the Gulf Stream because it flows out of the Gulf of Mexico. The current flows northward through the Florida Straits between The Bahamas and Florida at an average speed of four miles per hour (6.4 kilometers per hour). Navigators aboard Spanish treasure ships used this northerly current to their advantage during their return voyages to Spain.

On the eastern side of the Lucayan Islands are several main maritime passages such as the Mayaguana Passage, the Crooked Island Passage, the Mira Por Vos Passage, and the Turks Passage (or the Columbus Passage) in the Turks and Caicos Islands. Also bordering the entire Lucayan Islands' eastern seaboard is the western boundary of the Bermuda Triangle (or Devil's Triangle), with San Salvador Island jutting deepest into the Devil Triangle's western border. The triangle is bounded by an imaginary line connecting Miami, Florida, San Juan, Puerto Rico, and Bermuda in the North Atlantic Ocean in a triangular shape. The western boundary of this triangle runs along the Lucayan Islands' eastern seaboard.

The closest inhabited Bahamian island to the Turks and Caicos Islands is Great Inagua, approximately 40 miles (64 km) west of Providenciales Island. Inagua and Providenciales Islands are about 70 miles (112 km) from Hispaniola and 60 miles (96 km) northeast of Cuba's[65] southeastern coast.

The Turks and Caicos Islands has over 40 small islands and cays, with eight islands inhabited. The islands are approximately 575 miles (925 km) southeast of Miami, Florida, and 350 miles (563 km) southeast of Nassau, Bahamas. This British Overseas Territory is divided into two main island groups separated by a 20-mile (32 km) submarine trench named the Turks Passage, also called the Columbus Passage. The two groups consist of the smaller Turks group of islands on the Turks Passage's eastern side and the larger Caicos group on the Passage's western side.[66] The Caicos group is on the outer fringes of the expansive Caicos Bank.

The Turks group comprises Grand Turk at its northern end, where the country's capital (Cockburn Town) is located. Salt Cay is at the group's southern end. The two-mile Cay is approximately 10 miles (17 km) south of Grand Turk. The Caicos group consists of six main islands—South Caicos, East Caicos, Middle (or Grand) Caicos, North Caicos, Providenciales (also known as Provo), and West Caicos.[67] These islands are on the outer fringes of the Caicos Bank. Grand Turk and Salt Cay were known as the Turks Islands. The Caicos Islands were not included in this group until 1848, when they became known as the Turks and Caicos Islands. [68]

Brackish lakes, tidal creeks, and blue holes (deep inland or coastal sinkholes containing saltwater with freshwater lenses) are found throughout the islands. The most concentrated area of blue holes "on earth" is located on Andros Island in The Bahamas—approximately 175 blue holes inland and 50 offshore. [69]

There are three climatic zones within the Lucayan chain. These zones are determined by rainfall levels and temperature ranges that are conducive to specific types of vegetation. These zones are classified as the moist subtropical, moist tropical, and dry tropical zones.[70]

The moist subtropical zone consists of Abaco, Andros, Grand Bahama, and New Providence Islands, where most native pine trees grow. The moist tropical zone comprises the central islands, including Acklins Island, Cat Island, Crooked Island, Eleuthera, Exumas, Long Island, Rum Cay, Samana Cay, and San Salvador. The dry tropical zone includes Great Inagua, Little Inagua, Mayaguana, and the Turks & Caicos Islands.[71]

The climate is generally mild throughout the year. It is most appealing to visitors from colder climes, with the average temperature between the low 70s F (about 21°C) during winter months and low 80s F (about 27°C) during the summer months. The temperature seldom falls below the 60s F (about 16°C) or above the lower 90s F (about 32°C).[72]

The average annual rainfall is about 44 inches (1,120 mm) in The Bahamas, mainly in summer.[73] In the Turks and Caicos Islands, the annual rainfall averages about 29 inches (736 mm).[74] Prevailing winds come out of the northeast in winter and the southeast in summer, cooling a usually humid environment.

The annual hurricane (cyclone) season is between June and November.[75] Several of the islands are periodically hit by tropical cyclones and occasionally sustain extensive damages during the passage of a hurricane. Abaco and Grand Bahama in the northern Bahamas were struck in 2019 by Hurricane Dorian. Dorian was a Category 5 storm—the worst Atlantic storm in recent history, with winds over 200 miles per hour and storm surges over 20 feet.

Although others believe the numbers are higher, approximately 70 people were officially reported losing their lives during the storm. A severe storm that struck the Turks and Caicos Islands was Category 4 Hurricane Ike in 2008, leaving much flooding and damage to buildings.[76]

The natural vegetation found on these islands is Pine forests (mainly in the northern islands of The Bahamas), coppice (including Mahogany, Brasiletto, and Cacti plants), and mangroves.[77] Among the floras unique to the archipelago is the endangered Zamia Lucayana, a plant species in the Zamiaceae family. The plant is endemic to Long Island in the central Bahamas and is globally rare.[78]

Land faunas include iguanas and the hutia (a small rat-like rodent[79] eaten by Lucayans), a thriving West Indian Flamingo colony (the largest in the region) on Great Inagua Island in the southern Bahamas, and the Bahama Parrot. San Salvador Island is home to the Cyclura rileyi, otherwise known as the Bahamian rock iguana or the San Salvador rock iguana. Today, this iguana is an endangered lizard species in the family Iguanidae.[80] Flamingos and Iguanas are also found in the Turks and Caicos Islands.

The marine environment is filled with numerous marine faunas. Marine animals include dolphins, big game fish (tuna, Blue Marlins, and Mahi Mahi), whales (notably, sperm and humpback whales), snappers, groupers, spiny lobsters, and conch (a large sea mollusk or sea snail that is a native delicacy).

Interestingly, the Atlantis Resort Blue Project in The Bahamas sponsored the tracking of a marine leatherback turtle through the Caribbean and the east coast of the United States. The turtle is named "Lucaya."[81]

Lucayan Island Names

The Bahamas and Turks and Caicos Islands initially had Lucayan names. Early Spanish explorers and cartographers gave several of these islands Spanish names. Today, approximately six of the Lucayan names (or their derivatives) are still in use. Several of the smaller cays and islands have Spanish names. Early English settlers gave the remaining islands English names. Below is a list of Spanish Lucayan and English names for some of the main Lucayan Islands.[82]

Spanish Names	Modern Names	Taíno Names	Meaning
Inagua (Bahamas)	Inagua	i+na+wa	Small Eastern Land
Baneque (Bahamas)	Inagua	ba+ne+ke	Big Water Island
Guanahaní (Bahamas)	Little Inagua	wa+na+ha+ni	Small Upper Waters

Spanish Names	Modern Names	Taíno Names	Meaning
			Land
Utiaquia (Bahamas)	Ragged Island	huti+ya+kaya	Western Hutia Island
Jume(n)to (Bahamas)	Crooked/Jumento	ha+wo+ma+te	Upper Land of the Middle Distance
Curateo (Bahamas)	Exuma	ko+ra+te+wo	Outer Far Distant Land
Guaratía (Bahamas)	Exuma	wa+ra+te+ya	Far Distant Land
Babueca (Turks and Caicos)	Turks Bank	ba+we+ka	Large Northern Basin
Canamani (Turks and Caicos)	Salt Cay	ka+na+ma+ni	Small Northern Mid-Waters
Cacumani (Turks and Caicos)	Salt Cay	ka+ko+ma+ni	Mid-Waters Northern Outlier
Amuana (Turks and Caicos)	Grand Turk	aba+wa+na	First Small Land

Spanish Names	**Modern Names**	**Taíno Names**	**Meaning**
Caciba (Turks and Caicos)	South Caicos	ka+siba	Northern Rocky
Guana (Turks and Caicos)	East Caicos	wa+na	Small Country
Aniana (Turks and Caicos)	Middle Caicos	a+ni+ya+na	Small Far Waters
Caicos (Turks and Caicos)	North Caicos	ka+i+ko	Nearby Northern Outlier
Yucanacan (Turks and Caicos)	Providenciales	yuka+na+ka	The Peoples Small Northern [Land]
Ianicana (Turks and Caicos)	Providenciales	ya+ni+ka+na	Far Waters Smaller [Land]
Macubiza (Turks and Caicos)	West Caicos	ma+ko+bi+sa	Mid Unsettled Outlier
Mayaguana (Bahamas)	Mayaguana	ma+ya+wa+na	Lesser Midwestern

Spanish Names	Modern Names	Taíno Names	Meaning
			Land
Yabaque (Bahamas)	Acklins Island	ya+ba+ke	Large Western Land
Samana (Bahamas)	Samana	sa+ma+na	Small Middle Forest
Yuma (Bahamas)	Long Island	yu+ma	Higher Middle
Manigua (Bahamas)	Rum Cay	ma+ni+wa	Mid Waters Land
Guanahaní (Bahamas)	San Salvador	wa+na+ha+hi	Small Upper Waters Land
Guateo (Bahamas)	Little San Salvador	wa+te+yo	Toward the Distant Land
Guanima (Bahamas)	Cat Island	wa+ni+ma	Middle Waters Land
Nema (Bahamas)	New Providence	ne+ma	Middle Waters
Ciguateo	Eleuthera	siba+te+wo	Distant Rocky Place

Spanish Names	**Modern Names**	**Taíno Names**	**Meaning**
(Bahamas)			
Lucayoneque (Bahamas)	Great Abaco	luka+ya+ne+ke	The People's Distant Waters Land
Bahama (Bahamas)	Grand Bahama	ba+ha+ma	Large Upper Middle [Land]
Habacoa (Bahamas)	Andros	ha+ba+ko+wa	Large Upper Outlier Land
Bimini (Bahamas)	Bimini	bimini	The Twins

4

The Bahamas Resettled

Although the Slave Trade was abolished in British colonies in 1807, the practice of slavery was not abolished until 1834, followed by a four-year apprenticeship program. Despite several slave uprisings in the Bahama Islands, the relatively peaceful nature of Bahamians was evident immediately following the pronouncement of complete emancipation for slaves in 1838. The Bahama Colony's British Governor Francis Cockburn reported that there was not the "slightest disposition of tumult or insubordination" on the day (1 August 1838) the African slaves received their complete freedom.[83]

The native Lucayans (or, more precisely, their ancestors) inhabited the Bahama Islands since the 700s AD and were later claimed by the Spanish in 1492. About 45 years after the Lucayan extinction, it is believed the French Huguenots were the first Europeans to set sail for the "Lucayes"[84] (the French name for the islands) in an attempt to settle them in 1565.

The Huguenots were French Protestants in a Catholic country and wanted more religious, economic, and social freedom.[85] They had planned to settle on Abaco Island, which they called "Lucayoneque" at the time. However, there is no evidence of them ever arriving at the island, nor is it known what happened to them.[86]

England awarded the uninhabited Lucayan Islands to Sir Robert Heath in 1629,[87] although Spain never formally relinquished the islands until 150 years later. The area between 31 and 36 degrees north latitudes (generally North Florida to North Carolina and the islands to the south of Florida) was granted to Attorney General Sir Robert Heath in 1629. Heath was born in England and became a member of parliament before being appointed Attorney General under King Charles I of England.[88]

In addition to the land granted on the North American coast, the Attorney General was awarded the "...islands of Veajus (possibly Abaco) and Bahamas (Grand Bahama), and all other islands lying southerly or near upon said continent...." It is unclear if this grant included the Turks and Caicos Islands, as they would have been among the "other islands lying southerly...." Nonetheless, there is no record of any effort being made by Sir Robert Heath to settle the islands.[89]

Consequently, the Bahama Islands were granted to English colonists from Bermuda in 1647, who began resettling them in 1648, almost 20 years after Sir Robert Heath's grant was awarded. (Bermuda was initially called "Sommers Island" after Sir George Sommers from England, whose ship ran aground there in 1609).

Religious Puritans from Bermuda wanted to settle the Bahama Islands to pursue peace, economic and religious freedom.[90] The ruling class in Bermuda was loyal to King Charles I until his death in 1649 and later to his son, King Charles II (who was eventually proclaimed king by the Parliament of Scotland in 1649). Both Kings were heads of the Church of England.

Photo 6: Statue of a Puritan in Springfield, Massachusetts by American Sculptor Augustus Saint-Gaudens (Source: Wikipedia)

The Puritans, a religious branch within the Protestant movement, were discontented with England's religious practices. Unlike the English Pilgrims, another offshoot of the English Protestant movement that founded the Plymouth Colony in New England (U.S.) in 1620, the Puritans wanted to reform the Church of England from within the movement.[91]

The more radical Pilgrims wanted complete separation from the Church of England. Unlike the Catholic Church, the Pilgrims and the Puritans worshipped Jesus Christ as head of the Christian church and did not accept the Catholic Pope or the British monarch as heads of their respective faiths. The discord was a reflection of the political and religious changes in England at the time.[92]

The tension eventually erupted into Britain's first civil war, which began in 1642. Before this, King Charles I had become unpopular after dissolving Parliament and ruling by decree. This war evolved into several conflicts that were rooted in religious disputes. The civil wars involved England's Parliamentarians (or Roundheads) fighting against King Charles I's Cavaliers (Royalists). Roundheads and Cavaliers were demeaning names opposing sides called each other at the time. The term Roundhead was derived from how English apprentices shaved their heads. Cavalier referred to the Spanish "Caballaros," meaning armed troopers or horsemen.[93] Charles I was eventually defeated and beheaded in 1649.[94]

As Royalists, the Bermudan colonial administrators were less tolerant of those who resisted England's established political and religious order.[95] Captain William Sayle and others led the endeavor to form the "Company of Adventurers[96] for the Plantation of the Islands of Eleutheria formerly called Buhama in America, and the Adjacent Islands" in 1647.[97] (Eleutheria is now spelled Eleuthera, and the Company of adventurers is called the Eleutheran Adventurers). This group rented the islands from the British Crown.[98]

In 1648, Sayle and seventy English settlers from Bermuda made landfall at the northern end of an island in the central Bahamas, which they named Eleuthera, meaning "Freedom" in Greek. The new colonists included 28 enslaved persons of African descent.[99] The arrival of the Eleutheran Adventurers, more than a century after the Lucayan extinction, made Eleuthera Island the birthplace of the Bahamas.

The original Lucayan name for Eleuthera was Segatoo[100] (sometimes spelled Cigatoo).[101] The Eleutheran Adventurers' efforts were met with hard times from the onset. Just before landfall, one of the two ships (*The William*) struck a reef (now called Devil's Backbone) off mainland Eleuthera's north shore, east of Spanish Wells Island. Initially, the settlers lived in a cave known as Preacher's Cave, not far from their landing site. Preacher's Cave was described as the Plymouth Rock of The Bahamas.[102]

In 1649, the Stuart Monarch King Charles II (King Charles I's successor) was proclaimed King by Scotland's Parliament. He was later deposed in 1651. King Charles II went into exile in France during his deposition from 1651 to 1660, known as the English Commonwealth or the Interregnum.[103] His fleeing to France in 1651 effectively ended the war and allowed Oliver Cromwell to become Lord Protector of England.[104]

During Cromwell's installment as Lord Protector, England became a Republic and was without a ruling monarch. Although a Calvinist, Cromwell had a more favorable disposition toward the Puritans[105] who practiced their religious beliefs without persecution.

Photo 7: Preacher's Cave, North Eleuthera, The Bahamas. (Source: Tate A. Bethel)

In Eleuthera, the settlers eked a living from the little fruit and animals they could find or hunt on land and sea. Were it not for the warmer tropical climate, early Eleutheran settlers may have had a worst outcome than the English Puritan settlers at the Plymouth Colony in Massachusetts. Although nearly half the original settlers died within months of their arrival in Massachusetts in 1620, the Plymouth settlers received vital support from the local Indians.[106] The Plymouth Colony was the first American Puritan colony founded in 1620, 28 years before the Eleutheran Puritan colony was established in 1648.

Unlike the Plymouth experience, there were no surviving Lucayans to show the Eleutherans how to catch and prepare wildlife and survive off indigenous plants, such as the hutia (a small rodent), iguanas, flamingos, conch (a large sea snail or mollusk), and the cassava (manioc) plant (a woody shrub and a Lucayan staple). Survival had to be learned first-hand.

Although settlers received emergency provisions from the Massachusetts colony, most returned to Bermuda or the American colonies in North America within a decade. Many of the original Bermudan colonists, including the Bahama Colony's first proprietary governor and leader of the expedition, Captain William Sayle, left within ten years due to their hardships.[107]

The Bermudans' early venture coincided with the beginnings of buccaneering and piracy in the Caribbean region. The "vermin's" notorious activities spilled onto the shores of the fledgling Bahama Colony. The illegal loot was appealing to settlers who fell on hard times in the islands. The early English settlers' subsequent involvement in wrecking, piracy, and salvaging Spanish ships resulted in devastating attacks on Nassau and neighboring islands by Spanish and French naval forces.[108]

Livelihood for the settlers was sustained by trade between the British colonies in America and the British West Indies. Eleuthera provided brazilwood, ambergris (a whale secretion), seal oil, and goods salvaged from shipwrecks.[109] On one occasion, the Massachusetts colonists had shipped provisions to the Bahama Colony (Eleutheran Adventurers) to provide "succor" for the distressed colonists in 1650. The Bahama colonists exported ten tons of Brazilwood to Boston to help the Massachusetts colonists raise money for Harvard College as an expression of gratitude.[110] William Sayle and his family left Eleuthera for Bermuda in 1657.[111] He was reappointed Governor of Bermuda until 1662.

King Charles II returned from exile and was restored in 1660, shortly before Sayles reappointment as governor.[112] King Charles II reigned as King of England, Ireland, and Scotland from 1660 until he died in 1685.[113] In 1663, he granted the Carolinas to eight Lords Proprietors, who later appointed William Sayle as Governor in 1669.[114]

The Bahama Colony became a "dumping ground" for religious dissidents, troublesome Bermudans, slaves and free blacks, and Bermuda criminals during the colony's early years The territory had become a Little Australia, as Australia in the Pacific region was initially settled by criminals and exiles.[115]

Despite the challenges, more settlers arrived at the young Bahama Colony. The influx was sparked, in part, by the Bermudan ruling class declaring themselves Loyalists committed to King Charles II, and by extension, the Church of England after King Charles was exiled to France.[116]

Settlers began inhabiting other neighboring *Family Islands* (as islands in the Bahama chain are traditionally called by locals). William Sayle found refuge on the island of New Providence after being shipwrecked there and named it "Providence" in 1666. A new wave of Bermudans also began settling in New Providence that year.[117]

The newcomers called the island Sayle's Island. They later renamed it New Providence to distinguish it from Old Providence Island off the Central American coast (between Costa Rica and Honduras) that Puritans once settled and is now under Colombia's control.[118]

Under Sayle's influence, six Lords Proprietors from the Carolina Colony in North America became interested in the Bahama Colony and were granted it by King Charles II.[119] They assumed responsibility for governing the colony in 1670[120] but did not reside in the colony.[121]Sayle died in 1671. New Providence had a safe harbor and was the most populous island at the time. The 21-mile long (33 km) island became the seat of government.[122]

Nassau, New Providence's main settlement, was initially named Charles Town after King Charles II. In 1695, Charles Town was renamed Nassau after Prince William III of the House of Orange-Nassau in the Netherlands. Prince William was Dutch; he was born in the Netherlands and was of the House of Orange-Nassau. He was the grandson of King Charles I (who was beheaded in 1649) and the nephew of King Charles II (who fled to France the same year).[123]

Prince William had married the daughter of Charles II's brother James, the then Duke of York, who became the King of England and Ireland (as James II) and Scotland (as James VII) in 1685. Prince William became popular with European Protestants, especially in England. His father-in-law, King James, was Catholic and disliked by the Protestant majority in Britain. With the support of Britain's religious and political leaders, Prince William invaded England. He dethroned his father-in-law, King James II, in 1688, during the war known as the Glorious Revolution. [124]

Prince William of Orange-Nassau (son-in-law of King James) and his wife Mary (daughter of King James) became King and Queen of England, Ireland, and Scotland in 1689. William reigned until he died in 1702.[125] The naming of Nassau town by English colonists is perhaps the only unique Dutch influence King William III might have had on the Bahama Colony; albeit, indirectly. Cultural developments during that period reveal significant signs of English influence in such areas as language, architecture, law, and religion; much of this influence remains evident today.

Coincidentally, the first Dutch settlement in the Americas was named Fort Nassau. The Fort was established as a fur trading outpost on Castle Island near Albany in 1614 or 1615.[126] Today, Nassau is also the name of a village in Nassau Town, New York.[127] Nassau Town was settled in 1760 and was once named "Town of Philipstown" in 1806. The name was changed to "Nassau" in 1808.[128]

Illustration 3: - Map of the Bahama Islands (Source: Wikimedia)

From 1670 through 1704,[129] the proprietary governors managed the Bahama Colony affairs on behalf of the Lords Proprietors in Carolina.[130] The little participation in the Bahama Colony's development by the Lords Proprietors encouraged local governors to look after their personal interests with settlers seeking fast and easy means of providing for themselves. Hugh Wentworth was the first proprietary governor in the colony.[131]

Although a blight on The Bahama's beginnings, piracy was a common means of illicitly satisfying the colonist's needs during the infamous Golden Age of Piracy. From 1670 to 1718, The Bahamas and Turks and Caicos Islands went through a series of corrupt proprietary governors who cared little for the islands' development or their inhabitant's wellbeing. At times, the colony was devoid of any governorship. As a result of the lack of headship, Nassau was overtaken by tyrannical pirates, which provoked attacks on the colony by Spanish and French naval forces.

The governors' improprieties, coupled with French and Spanish retaliations, necessitated the British Crown taking the Bahama Colony's civil and military government under its control. The Bahama Islands were subsequently leased to the colony's first Royal Governor, Captain Woodes Rogers, in 1717. Captain Rogers took up office in 1718 (to 1721). His zero tolerance for pirates resulted in the sea bandits fleeing the islands for safer seas.

Rogers drove the last rogue pirates out of the Bahama Colony, including Edward Teach (Alias Blackbeard), Benjamin Hornigold, Calico Jack Rackham, Anne Bonny, and Mary Read. Those who persisted in the infamous trade were hung on the waterfront at Fort Nassau in 1718, where the renowned British Colonial Hotel is located in downtown Nassau.[132]

In 1728 (to 1731), during Woodes Rogers's second term as Governor, the British Privy Council authorized a general assembly for the colony. The assembly was a lawmaking body consisting of a Council appointed by the governor and the elected representatives of their respective island communities. Captain Rogers convened an assembly on 29 September 1729.

The Turks and Caicos Islands was authorized to send a representative but resisted doing so.[133] The Turks and Caicos Islanders wanted to remain strictly accountable to Bermuda to maintain their southern islands and salt industry control. The Lords Proprietors eventually gave up their remaining rights over the colony to the Crown in 1787 for £12,000.[134]

Adapting to island life was still a struggle for many a century after colonization began. Of the 100,000 loyalists who fled the American colonies following the American Revolution in 1783, 6,000 migrated to The Bahamas. With them came the hopes of thriving plantation economies. Unlike its British Caribbean Island counterparts of Jamaica, Barbados, and Trinidad and Tobago, the Lucayan Islands became a non-plantation colony due to unproductive soil and insect infestation, among other challenges.[135] Many post-American Revolution immigrants to the Bahama Colony returned to America or migrated elsewhere because of hardships experienced in the islands.

Some political progress was made during this period. By the time Loyalists arrived in 1784 from the former British colonies in America, the Bahama Colony was already practicing representative government. New Providence, Abaco, Andros, Harbour Island and Eleuthera, Exuma, and Long Island were all represented.[136] Free blacks and people of color were not given the right to vote until 1830 and then only if they were born free and not of African birth.[137] In 1833, the first black men were elected to parliament.

Between the pirate era and political autonomy, the colony's economic mainstay was sustained by opportunistic smuggling. Covert maritime activities included wrecking during the piracy era (1650 – 1718), arms and ammunition smuggling to the U.S. southern states during the American Revolution (1775 – 1783), and Civil War (1861 – 1865), and rum-running during the American prohibition years (1919 – 1932). Legitimate industries comprised agricultural products (pineapple and vegetables), salt and sponge industries.

The sponge industry collapsed due to a microscopic fungus disease in the 1930s.[138] The pineapple industry declined because of international competition at the turn of the 19th century.[139] Cruise and overnight tourism began with Cunard Steamship passenger arrivals and the Royal Victoria Hotel's construction during the mid-1800s.[140]

With its new constitution's enactment in 1964, the Bahama Colony attained internal self-government under its first Premier, Sir Roland Symonette (a white Bahamian). Sir Roland was also the white minority government leader of the United Bahamian Party (UBP). In 1967, black majority rule was also achieved for the first time under Premier Sir Lynden Oscar Pindling (a Black Bahamian). Although there was a major riot (the Burma Road Riot in 1942) and a general strike (in 1958), the historic journey to political leadership for the black majority of Bahamians was termed "a quiet revolution."[141] (See the third book in this series, *The Bahamas and Turks and Caicos Islands—Peace Capital of the Americas* for further information on this political transition).

On 10 July 1973, 325 years after its resettling by British colonists from Bermuda, The Bahamas peacefully achieved independence from Great Britain with Lynden Oscar Pindling as the nation's first Prime Minister. As an independent country and a member of the British Commonwealth, The Bahamas comprises a parliamentary constitutional monarchy, with the British Monarch as head of state and a Bahamian Prime Minister as head of Government.[142] The three branches of Government are the Executive Branch (consisting of the Prime Minister and Cabinet Ministers), its bicameral legislature consisting of the two houses of parliament (the Senate and the House of Assembly), and the judiciary.

The Bahamas' population in 2010 was 350,000,[143] approximately ten times the population of the Turks and Caicos Islands, which had some 35,000 people during the same year.[144] Today's (Bahamas) population is estimated at 400,000.[145] The ex-pat community is relatively small, with most of them being from the United States, Canada, and England. Twenty-two of the main islands are inhabited, from Abaco and Grand Bahama Islands in the north and Inagua and Mayaguana Islands in the southern Bahamas.[146] The country's capital is located on New Providence Island in the central Bahamas.

Since the repopulation of these islands in 1648, The Bahamas' leading settlements (Nassau on New Providence Island and Freeport on Grand Bahama Island) were transformed from fishing villages and rural lands into modern metropolises. Today, tourism and offshore banking are the country's two main economic pillars. In 2019, The Bahamas attracted 7.2 million visitors to its shores.[147]

5

The Turks and Caicos Islands—Belonger Country

The Turks and Caicos Islands is located at the southeastern end of the Lucayan Island chain, to the east of Mayaguana and Inagua Islands in the southeastern Bahamas and 575 miles (925 km) southeast of Miami, Florida. The indigenous Taíno people from Hispaniola periodically visited these southern islands before settling them, making them an indigenous gateway to the Lucayan Islands around 700 AD.

Approximately thirty years after the British Bermudans initially settled the Bahamas in 1648, the Turks and Caicos Islands were seasonally visited by a separate Bermudan group beginning around 1678.[148] The British did not formally claim the Turks and Caicos Islands until almost a hundred years later, in 1766.[149]

Historically, salt was a primary resource that attracted Bermudans to the Turks and Caicos Islands seasonal. Salt raking became the main occupation of the early settlers on Salt Cay. The two-mile (3 km) long Salt Cay, south of Grand Turk, naturally produced an abundance of salt from evaporated saltwater, turning salt raking into an economic mainstay for several centuries. Salt was also made in South Caicos on the western side of the Turks Passage.[150]

The thriving industry (otherwise known as white gold) produced enormous profits for the Turks and Caicos Islanders, as the sugar

Photo 8: Lighthouse, Grand Turk, Turks and Caicos Islands (Source: Wikimedia)

industry did for plantations in the Caribbean.[151] The largest in the West Indies, the salt trade in the Turks and Caicos Islands flourished for approximately 300 years (the 1660s to 1960s) before collapsing due to regional competition and the islands' incapacity to expand acreage for increased production.[152]

The Bahama and Turks and Caicos Islands' inhabitants have had administrative differences rooted in the salt industry as British colonies. As the smaller of the two British territories within the Lucayan chain, the Turks and Caicos Islands came under the Bahama Colony's jurisdiction on several occasions. The initial period began after the British Government took possession of the Turks and Caicos Islands and placed them under the Bahama Colony's jurisdiction in 1766.

The decision was rooted in the Bahama Colony's interest in the salt industry. The initial administration period lasted for over 80 years, from 1766 to 1848.[153] The Turks and Caicos Islanders objected to this. Their islands were initially settled by a separate group of Bermudans in the 1670s, who came under the British Bermudan Governor's authority instead of the Bahama Colony's control.

Before the 1770s, the Turks and Caicos Islands were classified as common land, and anyone could settle them.[154] Consequently, the Bermudan settlers had control of the islands. The British Government was influenced by the Bahama Island Governor, William Shirley, to take possession of the islands and place them under the Bahama Colony's administration. The Bahama Colony intended to reap the benefits of the Bermudans' lucrative salt industry and enacted laws to reinforce its control over the industry.[155]

Also, in 1766, the British Government dispatched a Crown Agent (Andrew Symmer) to look after the King's affairs in the Turks and Caicos Islands.[156] The Crown Agent was responsible for reporting to the Bahama Colony's Government. Naturally, the Turks and Caicos Islanders considered it their right to exercise full autonomy over these islands, claiming that their islands were settled by Bermudans a generation before Symmer's time.

With salt being the primary source of subsistence, Turks and Caicos Islanders were reluctant to permit the Bahama Colony to tax its salt industry, a primary resource for these southeastern islands.[157] Nevertheless, the Bermudans' pleas to the British Government for Turks and Caicos Islanders to remain under Bermuda's jurisdiction fell on deaf ears. Their islands were subsequently left in the hands of the Bahama Colony.[158]

During the early salt years, pirate ships and attacks by rival European powers (French and Spanish) were a menace to the Turks and Caicos Islands.

The Islands' shallow waters and strategic location near the Old Bahama Channel and Spanish ports in Cuba made the islands a favorite pirate hang-out during the Golden Age of Piracy.[159] The Turks and Caicos Islands were attacked or seized by French forces in 1706, 1753, 1778, and 1782.[160] Spanish maritime forces also raided these islands occasionally.

In the late 1770s and 1780s, the American Revolution resulted in American loyalists (those loyal to the British Crown) migrating to the Turks and Caicos Islands with their slaves, adding to the cheap labor pool and the development of cotton plantations that were unsuccessful.[161]

It was not until a century later that the Turks and Caicos Islanders' wish for administrative independence from the Bahama Colony was granted. In 1848, the Turks and Caicos Islands was permitted self-governance under the supervision of the Jamaica Colony until 1874. According to the British West Indies Geographical Review, tax revenues from the salt industry were unable to sustain the colony. The Turks and Caicos Islands was subsequently declared a British Crown Colony[162] and made a dependency of the British Jamaica Colony from 1874 to 1962.[163]

While it is a commonly held view that the Turks and Caicos Islands was a dependency of the former British colonies of The

Bahamas and Jamaica, Dr. Carl Mills dispels this myth claiming the situation was the opposite. The former Turks and Caicos Islands' Minister of Education and author argues that the Bahama Colony, Jamaica Colony, and Britain depended on the salt industry revenues of the Turks and Caicos Islands.

Turks and Caicos Islanders exported large volumes of salt to British colonial America, Newfoundland (Canada), and Europe. The commodity was used as a vital food preservative throughout local communities and aboard vessels at sea. Interestingly enough, a quarter of the taxes gained from the industry supplemented the Bahama Colony's budget.[164]

Unfortunately, revenues earned by both colonies and Britain were not reinvested in the Turks and Caicos Islands' development.[165] Adding insult to injury, the Turks and Caicos Islands were "under-represented" in the Bahama Colony's parliament. The Bahama Colony usually made policy decisions concerning the Turks and Caicos Islands without Turks and Caicos Islanders' input.[166]

By 1964, the Turks and Caicos Islands' salt industry in Grand Turk and South Caicos had collapsed due to competition and the inability to expand. In 1974, salt production in Salt Cay folded.[167] A part of this competition came from Inagua in the southern Bahamas. Ironically, Turks and Caicos Islanders established the first settlement on Inagua and started salt-raking in 1803.

The Henagua Salt Pond Company was formed in 1848. The business was later purchased, expanded, and mechanized by American investors (the Erickson brothers) in the 1930s and sold to Morton Salt in 1954. Today, the Inagua salt industry is the second largest in North America, producing one million pounds (453,592 kg) of salt annually.[168] Dr. Mills indicates that Morton Salt was initially interested in investing in the Turks and Caicos Islands. It is believed that the Bahama Colony might have shifted

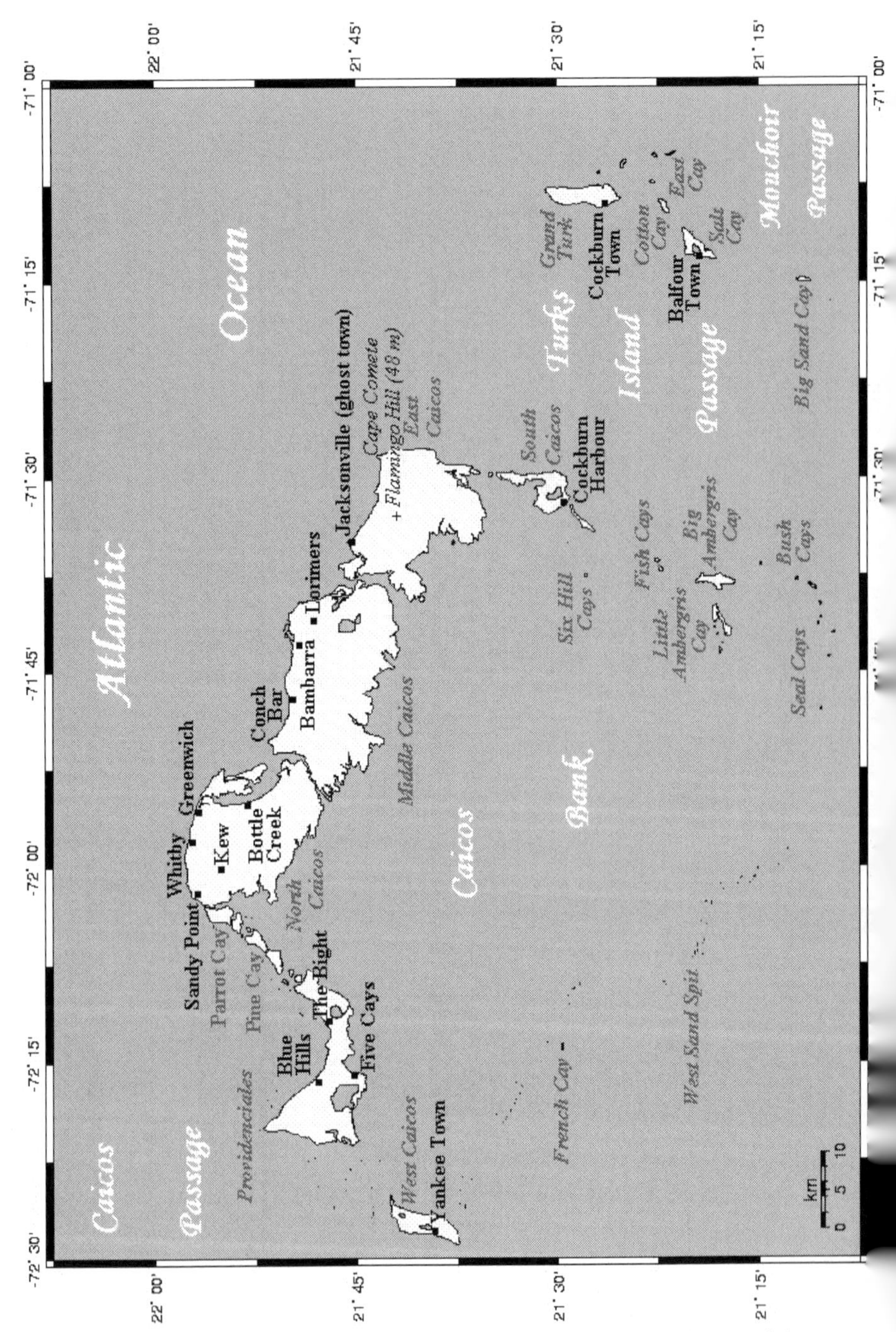

Illustration 4: Map of the Turks and Caicos Islands (Source: Wikimedia. By Kelsi)

the investor's interest in its capacity as the Turks and Caicos Islands' administrator, contributing to the industry's collapse.[169]

The Bahama Colony exercised jurisdiction over the Turks and Caicos Islands after Jamaica became independent in 1962.[170] On July 10, 1973, The Bahamas gained its independence from Great Britain. As a result, the Turks and Caicos Islands became a Crown colony, with the post of the administrator being elevated to governor. Although the Turks and Caicos Islands has maintained its current administrative status since The Bahamas' independence, Canada had expressed interest in making them part of its territory or annexing it as a province from as early as 1868. Talks of this union were held as late as 2014.[171]

Today, the Turks and Caicos Islands have an estimated population of 44,543 (2020),[172] with over 90 percent of African descent. Nine main islands and cays are inhabited, two in the Turks group (Grand Turk and Salt Cay) and six in the Caicos group (Providenciales, North Caicos, South Caicos, Middle Caicos, Pine Cay, Parrot Cay, and Ambergris Cay).[173]

Most of the islands' population is of the Christian faith, and eighty percent of the inhabitants live on Grand Turk, Providenciales, and South Caicos Islands.[174] According to a 2012 Census, approximately 38.9 percent of the population are citizens, and 34.7 percent are from Haiti. Other main demographic groups are the Dominican Republic (4.9%), United States (2.8%), Bahamas (1.8%), and United Kingdom (1.2%).[175]

Turks and Caicos Islanders often use the term "Belonger." This classification is applied to those who have citizenship rights in the Turks and Caicos Islands or Turks and Caicos Islander Status, previously known as Belongership. These rights include voting and holding government office and are only granted by the local government of the Turks and Caicos Islands. Belongers are generally associated with the Turks and Caicos Islands by birth or ancestry.[176]

Because the Turks and Caicos Islands is a British Overseas Territory, the United Kingdom awards British Overseas Territory Citizenship (BOTC) to qualified residents. Additionally, BOTC is the citizenship indicated on Turks and Caicos Islanders' passports. Most Belongers are identified as BOTCs.[177] Unlike persons with Turks and Caicos Islander Status, the BOTC status does not include citizenship rights such as voting.

The Islands' capital is on Grand Turk in the Turks group of islands. The capital carries the same name (Cockburn Town) as San Salvador Island's main settlement in The Bahamas—both settlements are named after the colonies' former Royal Governor, Sir Francis Cockburn. The Turks and Caicos Islands' seat of Government is also located in Cockburn Town, the territory's oldest settlement. Salt Cay is at the southern end of the Turks group of islands and was initially settled by Bermuda's salt rakers.[178]

Like The Bahamas, the Turks and Caicos Islands enjoys democratic governance. However, a British governor consistently administered this southeastern island group after the Bahama Colony gained independence in 1973. The Turks and Caicos Islands are now classified as a British Overseas Territory with internal self-government.[179]

The executive branch of its Government consists of the Governor-General, the Premier, and Cabinet Ministers. Unlike The Bahamas, its legislature is unicameral, consisting of the House of Assembly only.[180] The British monarch is head of state for the Turks and Caicos Islands (as an Overseas British Territory) and the Commonwealth of The Bahamas (as an independent nation). Similarly, tourism and finance are economic mainstays for both territories.

The first hotel (The Third Turtle Inn), marina (Turtle Cove Marina), and airstrip were constructed in Providenciales by 1970.[181] The cruise port in these islands is at the southern end of Grand Turk and was built in 2006 by Carnival Cruise Lines.[182]

Tourism arrivals exceeded 1.5 million for the Turks and Caicos Islands in 2019.

In 1976, Turks and Caicos Islander James Alexander George Smith McCartney (also known as J.A.G.S. McCartney) became the island territory's first Chief Minister. He served in this capacity from August of that year until his untimely death due to a plane crash in New Jersey in the United States in May 1980.

Before participating in his islands' political activities, J.A.G.S. McCartney had migrated to The Bahamas, where he campaigned on behalf of the Bahamian Progressive Liberal Party (PLP) during general elections in 1967. Lynden Pindling led the black majority PLP at the time. J.A.G.S. McCartney also organized several peaceful demonstrations for better working conditions in The Bahamas.[183]

There are a few interesting parallels between Pindling and McCartney. For example, Pindling's father was a Jamaican, and his mother, a Bahamian. The PLP leader became the Bahama Colony's first Prime Minister. J.A.G.S.'s father was also a Jamaican national, and his mother a Turks and Caicos Islander. Today, both men are hailed as national heroes.

Culturally, Junkanoo is a principal cultural festival shared by The Bahamas and the Turks and Caicos Islands. As early as the 1890s, some Turks and Caicos Islanders who returned from The Bahama Colony after seeking employment[184] helped establish the Junkanoo festival in their islands using knowledge and skills acquired in The Bahamas.

Also, J.A.G.S.'s political group in the Turks and Caicos Islands was formed out of a local youth group called the Junkanoo Club. Today, colorful Junkanoo parades with decorative costumes are the leading cultural festivals in The Bahamas and Turks and Caicos Islands.[185] The festivals are filled with rhythmic music created by goatskin drums' pulsating sounds accentuated with the vigorous shaking of cowbells and the blowing of conch shell horns.

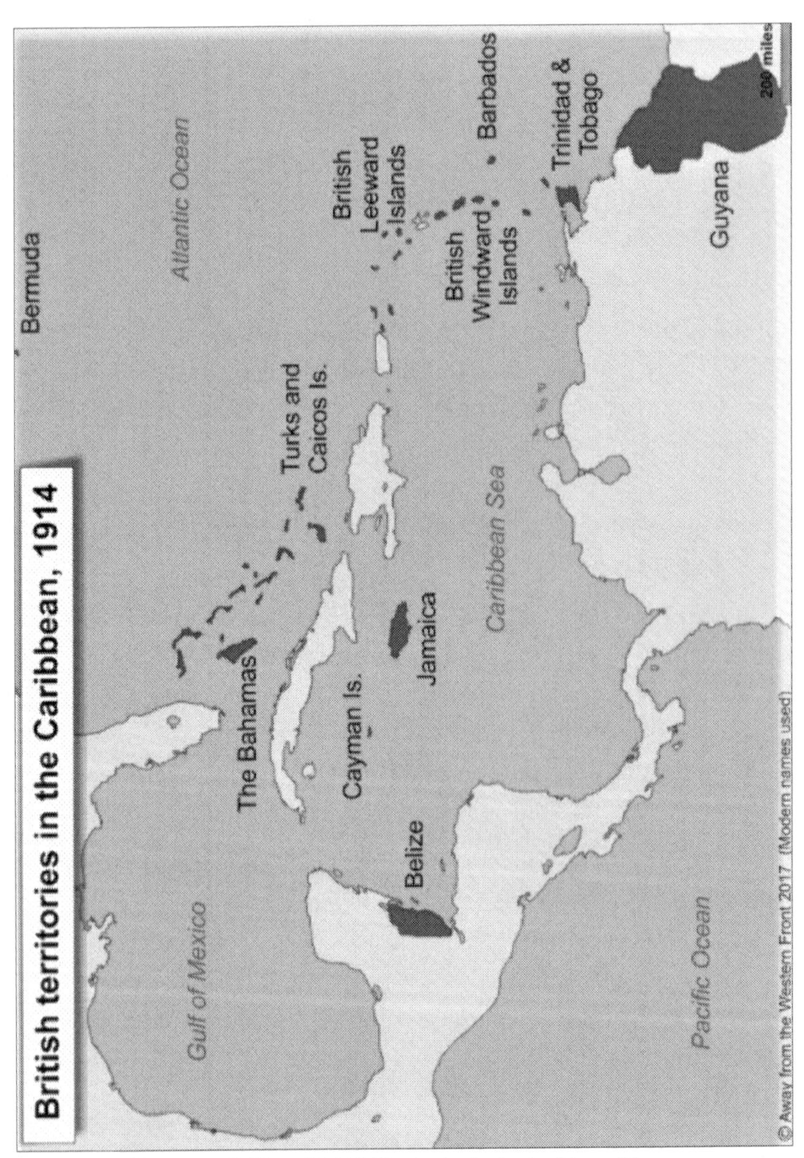

Illustration 5: Map of British Territories in the Caribbean (Source: www.awayfromthewesternfront.org)

Despite early administrative spats between the two colonies, many Turks and Caicos Islanders had migrated to the Bahama Colony to find employment during the colonial era, especially during the decline of their salt industry. Many were later employed in the Lumber industry in Grand Bahama during the first half of the 1900s.[186] Turks and Caicos Islanders are now the second-largest West Indian group in The Bahamas—the largest group are Haitian nationals.

Today, the Turks and Caicos Islands is a British overseas territory. Though politically separated, The Bahamas and Turks and Caicos Islands maintain a common bond of friendship and kinship. It is believed that the name this series proposes for the waters The Bahamas and Turks and Caicos Islands share as countries belonging to the Lucayan Archipelago would further strengthen this bond for the mutual benefit of both countries.

6

The West Indies—A Twisted Fate

The mistakes and fables of history are often found in names given to regions, territories, or ethnic groups. Sometimes, these errors and myths become more apparent after efforts are made to correct them. The names Antilles, Caribbean, and West Indies are regional examples of mistaken beliefs, yet they are vital links to understanding the past.

The names "Greater Antilles" and "Lesser Antilles" are appellations for the larger northern Caribbean Islands and smaller eastern Caribbean islands. These names were derived from a mythical island called "Antilia" featured on medieval charts. The island was thought to have existed in the Atlantic Ocean (west of Portugal), somewhere between the northwest coast of Africa and the East Indies.[187] Eventually, the name Antilia was applied to the Greater and the Lesser Antilles in the Caribbean Islands.[188]

Another example discussed earlier in this book is the name "Caribbean" from the name "Carib." This name was imposed on the indigenous people who dominated the region at the time of European contact. The name Carib is essentially a case of mistaken identity and is further explained in Chapter 9 titled, *The Caribs—A Mistaken Identity*.

Like the name Caribbean or Carib, the same can be said for the West Indies. This geopolitical region is also a case of mistaken identity. The name "West Indies" resulted from an attempt to correct Columbus' false belief that he had arrived in the East Indies. He subsequently called the people who greeted him "Indios" (Indians). [189] Columbus had accidentally arrived in the Ancient World of the Western Hemisphere, almost 10,000 miles (16,093 km) away from the East Indies' nearest lands, which included Cipangu (Japan), Cathay (China), and Indonesia.[190]

It was not until nine years after Columbus' first landfall in the Ancient World that Europeans realized he had discovered a world that was entirely new to him.[191]

This new world was located west of Europe and subsequently called the "Americas." It later became known as the West Indies to distinguish it from the East Indies.

Various European trading companies popularized the name West Indies during the 1700s and 1800s. For example, the Dutch West India Company, formed in 1621, had colonized territories in the Caribbean Islands, South America, and North America.[192] Also, the French West India Company was founded in 1664 with colonies in the Caribbean Islands, Central America, and South America.

Today, the name West Indies identifies the former or remaining European (overseas) territories in the Caribbean region (The Bahamas, Turks and Caicos Islands), the Caribbean Islands, and South and Central America. Consequently, British West Indies, Dutch West Indies, and French West Indies are names currently used to distinguish respective overseas territories in the Americas.[193]

It is also interesting to note that the Danish and the Swedes once had territories in the West Indies. In 1917, the Danish sold Saint Thomas, Saint John, and Saint Croix in the northern Lesser Antilles to the United States, which renamed them the U.S. Virgin Islands.[194] (The U.S. Virgin Islands are a small group of islands about 47 miles/77 km east of Puerto Rico. The British Virgin Islands are islands in the northeast portion of the group). The French ceded Saint-Barthélemy, an island at the north end of the Lesser Antilles, to Sweden in 1785. Sweden resold the island to France almost a century later, in 1878.[195]

The West Indies consists of two major archipelagos within the Caribbean region: the Caribbean Archipelago[196] and the Lucayan Archipelago. The Caribbean Archipelago is divided into two subregional island groups—the Greater Antilles and the Lesser Antilles.

The Greater Antilles are the much larger islands in the northern Caribbean chain, including Puerto Rico, Hispaniola (the Dominican Republic and the Republic of Haiti), the Republic of Cuba, Jamaica, and the smaller Cayman Islands.

The Lesser Antilles contain the remaining islands within the Caribbean Archipelago that form an arc-like shape starting from the Virgin Islands to the east of Puerto Rico in the northern Caribbean and continuing southward to the Republic of Trinidad and Tobago. Trinidad and Tobago is within 7 miles (or 11 km) of Venezuela's Caribbean coast in South America.

The Lesser Antilles is further divided into two sub-regional groups—the Leeward Islands that form the northern half of the Lesser Antilles and the Windward Islands that comprise the southern half. At the western extremities of the southern end of the Windward Islands are the Dutch (Netherlands) islands of Aruba, Bonaire, and Curaçao, north of the northwest Venezuelan coast.[197]

Illustration 6: Slaves planting sugar cane in Antigua (1823) (Source: Wikipedia)

States comprising the British West Indies were once a part of the British Empire. The Lucayan Islands (The Bahamas, the Turks and Caicos Islands) and Bermuda form the British West Indies' Atlantic sector.

The Lucayan Archipelago (The Bahamas and Turks and Caicos Islands) is north of the Caribbean Archipelago and is not bordered by the Caribbean Sea.[198] This archipelago is geographically separated from the Caribbean Islands. Its proximity to the Caribbean and close political and cultural affiliations make The Bahamas and Turks and Caicos Islands integral to the region.

Also north of the Caribbean Archipelago is Bermuda. This island is located in the middle of the North Atlantic Ocean. Bermuda is approximately 769 miles (1,237 km) southeast of New York or 900 miles (1,448 km) northeast of the Lucayan Islands. The tiny island is about 20 square miles (53 sq. km) in area with 71,000 people.[199]

Bermuda was named after the Spanish navigator Juan Bermudez, who discovered the island as early as 1503. The island was accidentally settled by the British as a result of a shipwreck in 1609. The English ship, Sea Venture, was in route to North America with supplies for the Jamestown Colony when the incident occurred. Today, Bermuda is a British Overseas Territory and is closely associated with the Caribbean region through its historical, political, and cultural ties with the British West Indies.[200]

Other Caribbean Islands that make up the British West Indies include Antigua and Barbuda, Barbados, British Virgin Islands, Cayman Islands, Dominica, Granada, Jamaica, St. Kitts and Nevis, St. Lucia, St. Vincent and the Grenadines, and Trinidad and Tobago. Guyana in South America and Belize in Central America are also a part of the British West Indies.

These examples of mistaken beliefs and attempts to correct them highlight the need to give the waters surrounding The Bahamas and Turks and Caicos Islands a name that represents the authentic heritage of these islands. It is believed that the name proposed by this series would help tell the origins of the region's history that is often overlooked.

Evolution of the Kingdom of England

The Kingdom of England, now called the United Kingdom of Great Britain and Northern Ireland, has had the most direct influence over The Bahamas and Turks and Caicos Islands since they were resettled after the Lucayan demise in 1648.

England was extensively involved with The Bahamas' and Turks and Caicos Islands' development for almost four centuries. The territory and official name for England have evolved over the years. What began as the Kingdom of England has since become the United Kingdom of Great Britain and Northern Ireland.

The name United Kingdom is also abbreviated as the U.K. or Britain. This name includes the island of Great Britain and its three countries—England, Scotland, and Wales—along with that portion of Ireland known as Northern Ireland. Northern Ireland shares its land border with the Republic of Ireland.

Historically, the "Kingdom of England" was formed by 927 AD,[201] after the union of the Anglo-Saxon kingdoms, including Northumbria, Wessex, and Mercia. Wales joined the Kingdom of England in 1536,[202] not long after the Lucayan Islands were stripped of the indigenous Lucayans by 1520. Scotland and England remained separate states until 1707.[203]

The Kingdom of England existed before and during the Age of Exploration or the Age of Discovery (15th through 17th centuries).[204] The first English colony within the Lucayan Islands was established under the Kingdom of England's jurisdiction in 1648. Other colonies formed within the Americas under the Kingdom of England's rule included Jamaica in the Caribbean and 12 of the 13 colonies founded on the east coast of today's United States of America from 1607 to 1732.[205]

Illustration 7: Map of The United Kingdom of Great Britain and Northern Ireland showing England, Wales, Scotland, and Northern Ireland (Source: Wikimedia. By Captain Blood).

In 1707, Scotland and England (including Wales) were united under "the Kingdom of Great Britain." This name lasted until 1801. The 13th American colony (Georgia) was founded during this period. Additionally, the Turks and Caicos Islands were claimed (1764) by the British, and the Bahama Colony's first Royal Governor, Captain Woodes Rogers, was appointed (1718) by the Kingdom of Great Britain. The American Revolutionary War (1775 - 1783) also took place during this time. The thirteen American colonies became known as British America before they won the American Revolution.

Later, in 1801, the union of Great Britain (England, Wales, and Scotland) with the Kingdom of Ireland became "the United Kingdom of Great Britain and Ireland," which lasted until 1922.[206] Ireland withdrew from the union with Great Britain, leaving Northern Ireland in the Union.

The name "United Kingdom of Great Britain and Northern Ireland" was later adopted. The Bahamas became independent, and the Turks and Caicos Islands became a British Overseas Territory under this final name. Queen Elizabeth II ascended the throne and became head of state for the two territories in 1953.

7

The Ancient World of the Western Hemisphere

Before their abandonment (around 1520) and later resettlement by Bermudan (English) Colonists in 1648, The Bahamas and Turks and Caicos Islands were inhabited by the Lucayans, whose ancestors originated from South America. According to available evidence, these Native Americans were the offshoot of an ancient migration that journeyed out of northeast Asia to seek a better life.

Interestingly enough, the root meaning of the word "tourism" indicates humanity's long-standing quest for a better life. From an etymological perspective, the Lucayan arrival made them the first "tourists" to these islands.

The word "tourist" comes from the word "tour,"[207] meaning "to travel about."[208] The word "travel" is derived from the word "travail," which initially meant "suffering."[209] In an original sense, a "tourist" is one on a quest for a better life that many call paradise. Ironically, the words explore (initially meant to weep or cry out), escape, vacation (from the word vacate), and paradise are key marketing words in today's tourism industry. Today, The Bahamas and the Turks and Caicos Islands are premier tourist destinations in the Americas that millions call paradise.

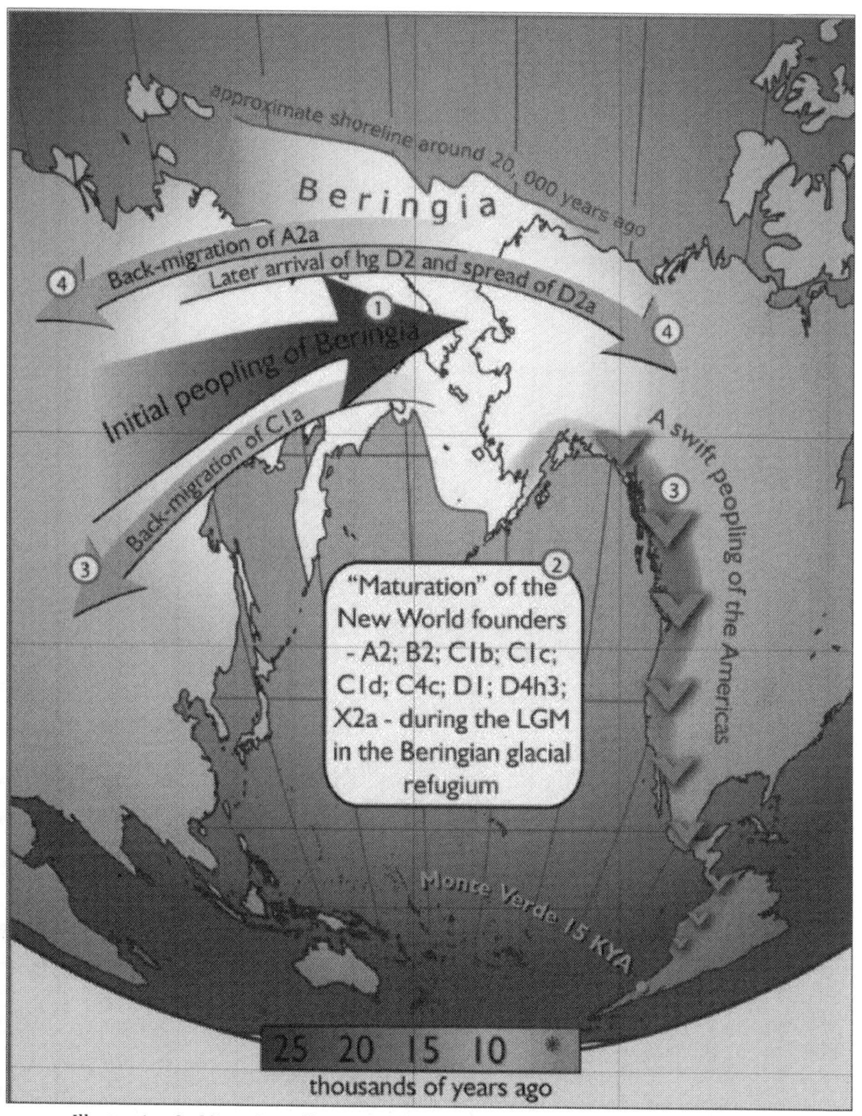

Illustration 8: Migration of human civilization out of Asia into the Ancient World of the Western Hemisphere (Source: Wikimedia)

The Lucayans are believed to be descendants of the Arawak-speaking people from South America who had migrated northward through the Caribbean Island chain. On their heels were another ethnic group out of South America considered more war-faring than the Arawaks. The Europeans called this latter group Caribs.

The Arawaks were subsequently driven northward in their quest for peace.[210] The ancient migration trek out of Asia eventually halted in the Lucayan Islands—where Paradise Island was later developed as a leading tourist destination. Except for the ice-covered Antarctica, the Western Hemisphere's continents were the last to be inhabited by modern humans.

The Lucayan arrival in the Lucayan Islands represented the end of the road for the Asian peoples who migrated to the Ancient World.

The Ancient World was once part of a supercontinent called Pangaea (or Pangea, both from a Greek word meaning "all the earth"). Pangaea was surrounded by one ocean known as Panthalassa. After the supercontinent began to break apart due to plate tectonics (or movement of the Earth's surface), the Ancient World was left partially attached to Asia before being entirely separated by water. The Asian continent remained permanently joined to the African and European continents.[211]

Before being separated by rising seawater, Asia was geologically attached to the Ancient World's northern continent by an exposed stretch of land between northeast Russia and northwest Alaska called Beringia. This land bridge enabled Asian peoples and animals to populate the Ancient World. The northwestern portion of the Ancient World's northern continent (including the Yukon area of western Canada and the western end of Alaska) became the gateway to (or the birthplace of) the Ancient World. Beringia disappeared under 98 to 164 feet (30 to 50 meters) of water[212] more than 11,000 years ago[213] as ice caps melted and sea levels rose.

The melting caps created a maritime space, now known as the Bering Strait, at least 53 miles (85 km) wide between Russia and the Ancient World. (The Bering Sea was named after Vitus Bering, a Danish sea captain, who Russia contracted to explore the area)[214]. The distance between the two continents is roughly equivalent to the maritime space between east Florida and the northern Lucayan Islands.

The rising sea later covered the land-bridge joining the continents, thereby separating the two continents. Essentially, the Western Hemisphere's continents became a lost world to the inhabitants of the Asian, African, and European continents (collectively known as the Old World).[215] Nevertheless, the Old World of Europe would reunite with the Ancient World thousands of years later and forcefully reshape it into a new one pursuing a better life.

Historically, several theories suggest how the Ancient World became inhabited. Some researchers argue that the Ancient World may have been populated by indigenous peoples from several points in Asia and the Pacific (including Australia).[216] Others hypothesize that people from southeast Europe may have settled the Ancient World by way of the Atlantic Ocean.[217]

The more credible theories indicate that human civilization first reached the Ancient World from East Asia, corresponding with archaeological finds and DNA data collected by National Geographic's Genographic Project. The Genographic Project is a research initiative that traces human genes' ancestry to determine human migration paths.[218]

Most population theories generally agree that the Western Hemisphere's continents were the last to be reached by the ancestors of today's human civilization. Genetic research also supports this idea.

The widely accepted land-bridge (Beringia) theory suggests that the ancient trek of human civilization began on the African continent and then made its way through Asia and Europe before crossing the Bering land bridge that once joined Russia and Alaska.[219] Ancient people then journeyed southward through North, Central, and South America.[220] One researcher estimated it took about 1,000 years for human civilization to travel from Alaska in the north to the southern end of South America.[221]

Recent research indicates that humans had initially settled both sides of Beringia some 20,000 years ago instead of the traditionally held view of 15,000 years ago. Research further suggests that those who might have arrived in boats had settled various places along the North American Pacific coast. [222]

Some scholars believe that Native Americans had settled the Caribbean[223] from points in North America (Florida),[224] Central America (Yucatan),[225] and South America (Venezuela) before the arrival of the Arawak-Taínos and Caribs in the Caribbean Islands and the Lucayans in The Bahamas and Turks and Caicos Islands.[226]

Others argue that Florida was populated with indigenous people approximately 14,000 years ago. However, there is no evidence of Native Americans migrating from Florida to the Lucayan Islands 50 miles (80 km) east of Florida.[227] Additionally, archeological evidence indicates that a branch of the ancient migration out of Asia was the first to make its way across the Yucatan Peninsula (in Central America) into the northern Caribbean Islands through Cuba as early as 5,000 BC.[228]

Other branches traveled farther south through Mesoamerica (the Central American region), eventually reaching the north coast of South America. From this area, offshoots of this ancient migration (including an Arawakan-speaking people) island-hopped northward along with the Caribbean Islands as early as 800 BC. The Caribs followed these ethnic groups in or around 600 AD,[229] 800 AD,[230] or 1200 AD.[231] It is also believed that the

Lucayan chain of islands was the last to be settled within the Caribbean region. The Lucayan Archipelago is located north of the Caribbean chain.

It is widely accepted that the Taínos of the Greater Antilles in the northern Caribbean, the Caribs of the Lesser Antilles in the eastern Caribbean, and the Lucayan Islands' inhabitants (the Lucayans) were among the larger groups of indigenous island peoples during European contact in the Caribbean region.

8

Populating the Caribbean—Arrival of Arawaks & Taínos

There are ongoing debates concerning the land and sea routes the Ancient World's indigenous peoples took to access the island chain now known as the Caribbean. The timing of their arrivals and their villages' locations are also being debated.[232] Throughout these discussions, old theories are either strengthened or disproved as new ones emerge due to new technologies that reveal the latest evidence.[233]

Nevertheless, scholars and researchers generally agree that multiple cultural groups initially colonized the Caribbean from the American mainland, which began several thousand years before European contact.[234] Two primary areas are widely accepted as main access points to the Caribbean during its original settling by indigenous peoples.

The areas are the Yucatan Peninsula in Central America and the northeastern coast of Venezuela in South America.[235] It is believed that there were two main migratory routes into the Caribbean from the continental mainland. More recent research suggests that there is a possible third entry area into the Caribbean. This area lies in the Isthmo-Colombian region, between northwestern Columbia and Panama.[236]

It is also theorized that cultural groups from the Central American mainland were the first to inhabit the Caribbean.

These groups lived in the northern Caribbean for about five thousand years before Europeans arrived. The other groups that colonized the southern and central Caribbean regions are believed to have arrived over 3,000 years later.

Scholars traditionally name cultural groups after the sites where archaeologists have found evidence of their existence.[237] Indigenous groups that inhabited the Caribbean before European arrival included the Casimiroid and Ostionoid peoples based on pottery and artifacts produced.

Some scholars propose that distinct cultural ethnicities emerged from these groups over time. For example, the Saladoid people from South America were ancestors of the Ostionoid people of Puerto Rico.[238] Also, the Ostionoid people were the ancestral group of the Taíno and Lucayan peoples.

Inevitably, various peoples and cultures became fully indigenous to the Caribbean region over extended periods. Those who came either co-existed, blended with other groups, or died out as new cultures emerged. Consequently, cultural groups such as the Saladoids and Casimiroids continued to evolve into new cultures within the Caribbean chain.

The indigenous peoples from Central America had settled the Greater Antilles for thousands of years before Columbus' arrival. On the other hand, the indigenous peoples from South America had inhabited the Lesser Antilles hundreds of years before European contact. This separation from the mainland naturally resulted in cultural and linguistic differences between the new groups and their ancestors.

The prevailing belief is that the Taínos and the Lucayans were descendants of the Arawak-speaking people from South America. The Arawaks initially inhabited South America's northeast coast near the Orinoco River in Venezuela. This group is believed to have migrated northward along the Caribbean Islands before reaching Puerto Rico.[239]

The word "Arawak" or "Aruac" means "eaters of meal."[240] Cassava was a main staple grounded into meal or flour to make flat Cassava cakes or cassava bread.[241] Arawaks came out of the Amazon basin region or from as far west as the Colombian Andes from among the Arhuaco and other Chibchan-speaking peoples in South America.[242] Knowledge of the Arawaks as a distinct indigenous group first came to light during the British exploration of the Guianas by Sir Walter Raleigh in 1595.[243] The migration of the Arawaks into the southern Caribbean took place around 800 BC to 200 BC.[244] Saladoid pottery was imported into the Caribbean from Venezuela during this period.[245] These Arawaks are believed to be associated with the culture of the Saladoid people of South America.

The Saladoid people were named after the archaeological site discovered at Saladero, near the Orinoco River's mouth in northeast Venezuela. Scholars have used their distinctive white-on-red pottery designs[246] to track their movements through the Caribbean.[247] The Saladoid people were also Arawak-speaking people[248] who made their way northward in dugout canoes (made from tree trunks).

The Saladoid migration into the Caribbean was followed by other ethnic groups arriving in the region over the next thousand years.[249] Among the groups were the widespread Arawak-speaking people.[250] However, a unique culture developed among the Saladoid people around 600 AD that archaeologists call Ostionoid. The Ostionans (from the Ostionoids in Puerto Rico) had also migrated to Hispaniola, from among whom the Taino Culture evolved,[251] developed around 1100[252] or 1200 AD.[253] The Ostionoid people were named after the Ostiones Site discovered in Puerto Rico.[254]

Illustration 9: Female Taíno Chief Anacoana born in Xaragua (Haiti) (Source: Wikimedia)

Several thousand years before the arrival of the Ostionoids in Puerto Rico and eastern Hispaniola (the island of the Republic of Haiti and the Dominican Republic), the Casimiroid people (a Stone Age people) had populated Cuba and Hispaniola during the Lithic Age, around 5000 to 4000 BC.

These ancient people had migrated from the Yucatan Peninsula in Central America into Cuba, reaching as far east as eastern Hispaniola in the northern Caribbean.[255] By the time Columbus arrived in Hispaniola in 1492, the Taíno culture was firmly rooted, with at least 400 years in the making from around 1100 AD.

Additionally, the Taíno people had already spread throughout the Greater Antilles' larger islands (Hispaniola, Puerto Rico, Cuba, and Jamaica). The Taínos had also inhabited several islands in the northern part of the Lesser Antilles. The islands included the Virgin Islands, St. Kitts and Nevis, and Antigua and Barbuda.

The name "Taíno" possibly came into use by the Taínos just before or around European contact.[256] Columbus was the first to use the name. Columbus mentioned the name "nitayno" in his diary after his initial encounter with the Taínos during his first voyage to the Ancient World. He was uncertain if the name referred to a Taíno noble, governor, or judge.[257]

Additionally, it is believed that the Taínos in Hispaniola had used the word "Taíno" during an armed conflict with the Spaniards to distinguish themselves as good people and not cannibals.[258] In 1836, this name was first used during an academic session involving the Taino language. The Taíno name has since been used to describe the ethnicity of these native people[259] and to distinguish the "island" Arawaks from the South American "mainland" Arawaks.[260]

Scholars have divided the Taínos within the northern Caribbean region into major regional groups. Those who inhabited Eastern Cuba, Hispaniola, and Puerto Rico are referred to as Classic

Taíno. Those of Central Cuba, Jamaica, and the Lucayan Islands are classified as Western Taíno. The Taínos of the Lesser Antilles's northernmost part (including the Virgin Islands) just east of Puerto Rico are called Eastern Taíno. Those who settled the Lucayan Islands are also described as Lucayan-Taíno.[261]

These broad-based classifications do not specify individual differences between ethnic groups in the northern Caribbean.[262] Although language and culture varied among Taíno groups,[263] the Taíno language appeared to be the most dominant.[264]

Hispaniola was host to one of the largest Taino groups, the Classic Taínos. The Classic Tainos were divided into five kingdoms on Hispaniola. The Classic Taínos of Puerto Rico, a smaller island, had 21 chiefdoms under a single confederated kingdom known as Borinquen.[265]

A Taíno chief, or "cacique, headed each dominion."[266] The term Cacique is derived from the Taíno word "kassiquan," meaning "to keep the house." The word was later translated as "prince" or "king" in English.[267] Women such as Anacoana also became Caciques.[268] The Caciques were responsible for warfare, political, and trade affairs. They had several wives. Upon a Cacique's death, his favorite wife was buried alive with her deceased husband's body.[269]

While the exact origins, timings, locations, and arrivals of the Ostionoids, Casimiroids, and others are still being debated, it is believed that the descendants of some of the groups melded and culturally evolved to form the Taíno culture.

The Taíno culture was eventually exported northward into the Lucayan Islands by Taíno immigrants who made cultural adaptations within their new environment. Here, in the Lucayan Islands, they became known as Lukku Cairi (now known as Lucayans).

9

The Caribs – A Mistaken Identity

The Caribs were among the last indigenous groups to colonize the Caribbean. Like the Arawaks, the island Caribs originated from South America's northern shores, where their ancestors called themselves "Kariña" or "Kalina."[270] According to archaeological research, the Island Caribs were also descendants of Arawakan-speaking people from northern South America.[271]

Mainland Caribs spoke Cariban, which belonged to the same broad Arawakan language as that of the Taínos of Hispaniola and the Arawaks on the north coast of South America.[272] The Cariban language family may have comprised up to three dozen languages, mainly around the northern Amazon area.[273] Although mainland Caribs spoke Cariban, the language of the island Caribs was different.

The Caribs arrived in the Lesser Antilles as early as 600 AD[274] or around 1200 AD.[275] It is believed that the Island Caribs island-hopped their way through the Lesser Antilles from South America. This ethnic group was considered more warlike than the mainland South American Caribs and the Taínos of the Greater Antilles.[276] The Caribs eventually occupied the Caribbean

Photo 9: Historical Mixed Media Figure of a Carib Warrior (Source: Wikimedia)

Islands from Trinidad at the Windward Islands' southern end to as far north as Guadeloupe, at the Windward Islands' northern end.

Some historians believe that the more war-faring Caribs displaced an earlier group of settlers in the Lesser Antilles known as the Igneri people. The Igneri people originated from the Arawak-speaking people of mainland South America. It is believed that the Caribs married the Igneri women.

Others suggest that the Igneri may have been the "terminal phase of the Arawak cultural development" just before European arrival in the southern Caribbean Islands.[277] During the 1450s, Caribs were launching raids on Puerto Rico and Hispaniola in the northern Caribbean chain.[278]

Ironically, the name Carib is of Taíno origin.[279] It is the name the Taínos of Hispaniola called their enemies from "the east."[280] In Taíno mythology, Caribs were cannibals. The Taíno word for cannibal was "Caribe." Europeans had no firsthand account of Caribs being cannibalistic predators or "New World savages."[281]

Columbus was the first European to mention the Caribs in his diary during his first voyage to the Americas.[282] He assumed the Caribs were cannibals based on the initial description given him by the Taínos in Hispaniola of their Carib raiders.[283]

The name Carib is a corruption of the Taíno words "cariba" and "caniba."[284] This name eventually became "Canib" and synonymous with the word "cannibal" or "man-eater."[285] Columbus had also confused the name Carib with Caniba, a people under the Great Khan of China's rule.[286]

The notion of Caribs being cannibals might have been reinforced after Columbus' men saw what appeared to be human bones in a Carib house in Guadeloupe by a Columbus' landing party. It is believed that the remains might have been associated with the ritual eating of relatives after they died.[287] Nonetheless, the unverified belief that Caribs were cannibals gave Columbus and his contemporaries the legal right to enslave them.[288] Hundreds of Caribs were captured and transported as slaves to Hispaniola.[289]

Columbus encountered the Caribs while exploring the islands they colonized in the Lesser Antilles during his second voyage in 1493. During the expedition, crewmembers freed several Taíno women on Guadeloupe Island, whom the Caribs captured in Puerto Rico. Just to the south of Guadeloupe, Columbus confronted Caribs on the island of Dominica during the same voyage.[290] The Carib militancy was immediately evident during these episodes. Their weapons of choice were the bow and arrows with poisoned arrows and clubs.[291]

Caribs were feared by their counterparts and were known to attack Taíno villages, killing the men and taking the women away as slave-wives. Raids on Taíno villages and their women's capture added to the influx of Arawakan speakers and culture in the Carib communities of the Lesser Antilles.[292]

Although the name Carib is said to mean "strong" or "brave" person,[293] the island Caribs originally called themselves "Kallinago" in their native men's language or "Kalliponam" in their native women's language.[294] The Language difference is attributed to the Carib men, who spoke Carib, capturing Arawak-speaking women.[295] Today's descendants of the so-called (Island) Carib people in the Commonwealth of Dominica refer to themselves *as* Kallinago (also spelled Kalinago).[296]

According to a French Franciscan, Father Raymond Breton, the island Caribs called themselves "Callinago" (spelled Kallinago). The women called the men Calliponam. Breton was a Dominican missionary and linguist from France who lived among the Kallinago people of Dominica in the French West Indies from 1641 to 1651. Breton also wrote two dictionaries translating French into Carib and Carib into French.[297]

There were also other name variations that the Caribs used: Calinago, Kalina, Karinaku, and Karina. Callinago possibly means "harmless," from the grammatical stem "kari" meaning "hurt" or "harm" and the suffix "na" meaning "(free) from." According to

Breton, "Callinago" means "Homme Paisible" (Peaceful Man) or "Homme de Bien" (Good Man).

Among the names used by the Carib enemies or the Europeans were: "Karifuna," Galibi," "Canibi," "Caribe," and "Carib." These names carry the connotation of "fearsome" or "hurtful."[298] The Spanish word for "Carib" was "Caribe," and the French equivalent was "Galibi."[299]

Europeans subsequently applied the name Carib to the islands in the region that were dominated by the Kalinagos (Caribs) and the waters these islands encompass. The name Carib was imposed on the Kallinagos by European "outsiders" with its cannibalistic implications.[300] Today, these islands, sea, and people are known as the Caribbean Islands, the Caribbean Sea, and the Caribbean people.

The Lesser Antilles' European occupation began during the first half of the 16th century, a period of intense colonial expansion. The Spanish had already occupied the Greater Antilles' larger islands—Puerto Rico, Hispaniola, and Cuba. Therefore, newcomers such as France and England were less likely to suffer retaliation from Spain for colonizing the islands in the eastern Caribbean (the Lesser Antilles).

These competing European powers were interested in taking full possession of fertile lands occupied by the Caribs. They wanted to establish elaborate plantations for growing cash crops such as cotton, sugar, and tobacco. The Europeans were acutely aware that the Caribs were determined to defend their islands fiercely.[301]

The Caribs' fierce opposition to colonial presence fueled European determination to extract or destroy them in these islands.[302] The Caribs offered so much resistance that the Carib-dominated region in the Lesser Antilles became known as the "poison arrow curtain"[303] by French and English forces.

Caribs were also known to attack European vessels in their waters, capturing Europeans and black slaves aboard. Ships that ran aground with passengers and human cargo were convenient targets.

The capture of black slaves by Caribs contributed to mixed African and Carib communities.[304] The Caribs also raided European settlements throughout the 16th and 17th centuries. European settlers were forced to live together with Caribs, and only through warfare were they able to take possession of Carib territories. As a result, Caribs did not make good slaves.[305]

There were also reports of Caribs uniting with runaway blacks in their fight against French and English settlers in the eastern Caribbean. For example, Governor Stapleton reported in 1667 that the Caribs and runaway blacks posed the most significant threat against English settlers in Antigua. This confrontation was also experienced in the French colony of Martinique.[306]

The Garifuna people are an offshoot of the Carib-African integration. This ethnic group exists today and is known as the Black Caribs. The Garifuna people were originally from Saint Vincent and the Grenadines in the eastern Caribbean.[307] The British won control of Saint Vincent and the Grenadines following an uprising of the islands' Garifuna people in 1797.[308]

That year, the last of the Garifuna people were deported by the British from their native homeland to the Spanish-controlled island of Roatán. Roatán is located off the east coast of Honduras in Central America. Today, the Garifuna people have well-established communities in Belize, Guatemala, Nicaragua, and Honduras in Central America.

Added to existing Carib communities are approximately 3,000 Kalinago (Carib) descendants living in the Commonwealth of

Dominica, an independent country.[309] Also, the Dominican government allocated 3,700 acres (15 km2) of land reservation, called Kalinago Territory (formerly Carib Territory or Carib Reserve), for Dominica's Carib population. In 2015, the Dominican government officially approved the name change for its Carib people and its territory. They are now known as Kalinagos and the Kalinago Territory.[310]

Descendants of indigenous Caribs also live in several countries near the north coast of South America. Among these countries are Venezuela, Guyana, Surinam, and French Guiana.[311]

Illustration 10: Artist rendition of a male and female Lucayan

10

The Lucayans—End of an Ancient Trek

***They seem to live in that golden world** of which the old writers speak so much, wherein men lived simply and innocently without enforcement of laws, without quarreling, judges and libels, content only to satisfy nature. - Peter Martyr,[312] a Portuguese historian employed by the Spanish Crown.*

Despite ongoing deliberations about the origins of indigenous peoples in the Caribbean, it is worthwhile mentioning that these debates do not detract from the consensus that the friendly people who welcomed Columbus to the Ancient World were an offshoot of an ethnic group that originated from the Greater Antilles (northern Caribbean).

Further, it is generally agreed that the Lucayan Islands were among the last islands inhabited within the Caribbean region before European contact.[313] The arrival of the Lucayans in the Lucayan Islands represented the end of an ancient migration mainly out of Asia. This migration had spread through North America, Central America (and the northern Caribbean), and onward through South America before continuing northward along the Caribbean chain.

The Lucayan ancestors were the first to inhabit the Lucayan Islands permanently. Here, they developed a distinct culture for over seven centuries before Columbus arrived in the Ancient World. This ethnic group also witnessed the beginnings of the

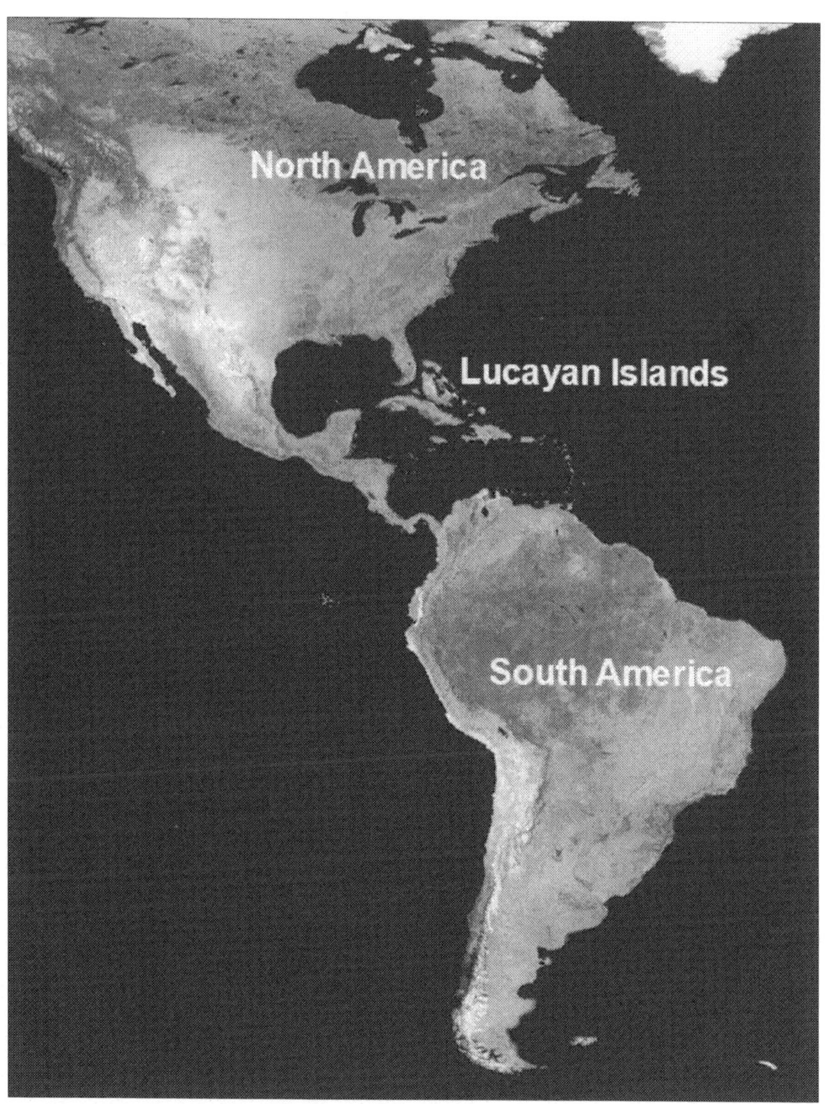

Photo 10: Map of the Americas (Source: Wikipedia. Note: Names were inserted on the original photo)

Old World's permanent reunion with the Ancient World, which later became the Americas. These people called themselves Island People in their native tongue (Lukku Cairi), indicating their cultural distinction. Today, they are known as Lucayans. Some scholars refer to the Lucayan people as Lucayan-Taíno.[314]

As previously mentioned in Chapter 3 of this book, the Anglicized name "Lucayan" is a historical blend of native Taino, Lucayan, Spanish (Latinized), and English (Anglicized) influences. Lukkunu Kairi is another Arawakan name the Lucayans might have called themselves.[315]

Lukku Cairi is said to mean "Island People." (Cairi or Kairi is Island, and Lukkunu or Lukku is People or Man}.[316] "Cairi" eventually became "cayo" in Spanish and "cay" in English. "Lucairi" is another word that means "Island People."[317]

Although the Lucayans became a distinct group, they maintained some cultural similarities with their Caribbean counterpart, the Tainos. Unfortunately, the Lucayans were the first indigenous group within the Ancient World to suffer extinction by European colonists at the beginning of a new era for the Ancient World of the Western Hemisphere.

Among the dominant indigenous groups that existed within the northern Caribbean region when Columbus made landfall in the Ancient World according to Spanish and French historical accounts were the Taíno, Carib, and Lucayan peoples.[318]

Two smaller ethnic groups in the region included the Ciboney people from central Cuba and the Guanahatabey people who occupied western Cuba. There is no firsthand account of this latter group.[319]

It is believed that the Ciboney people out of Cuba may have inhabited the Bahama Islands before the Lucayans. Little evidence has been found to confirm this belief.[320] Others argue that the Ciboney people were no longer a distinct ethnic group when the Europeans arrived in the Americas.[321]

Nevertheless, scholars posit that the settling of the Lucayan Islands took place during three overlapping periods. These timeframes are categorized as the Non-Lucayan Period (700 – 1300 AD), the Early Lucayan Period (700/800 – 1100 AD), and the Late Lucayan Period (1100 – 1530 AD).[322]

The Taíno culture in Hispaniola was not developed until after 1100 AD.[323] Consequently, the initial Lucayan Islands' settling began with the Taínos' ancestral group, the Ostionans (an offshoot of the Saladoids), from whom the Lucayan culture evolved. Archaeological evidence indicates that the Ostionans initially visited the Lucayan Islands in the south (Turks and Caicos Islands) from Hispaniola or Puerto Rico. The Ostionans formed the first wave of inhabitants in the Lucayan Islands at the beginning of the Non-Lucayan Period (705 - 1000 AD).[324]

It is also believed that the central Lucayan Islands of New Providence and San Salvador were settled by 800 AD during the Early Lucayan Period (700/800 - 1100 AD).[325] The presence of early Taínos significantly expanded among the Lucayan Islands during this period. Among these islands, the permanent settlers adapted to their new environment.

Some historians argue that the Taíno people from Hispaniola (and possibly Cuba)[326] began entering the Lucayan chain from farther north in the 1300s. The Taíno arrival in the northern islands occurred during the Late Lucayan Period (1100 - 1530 AD),[327] at which time the Taíno culture was well established.

Recent research using radiocarbon tests (of landscape disturbance) suggests Lucayans might have taken less than a century to spread throughout the Lucayan archipelago.[328] These tests were carried out in Abaco, The Bahamas' northernmost island.

The Lucayan culture had steadily evolved over 800 years from when the Ostionans' visited the islands in the 700s AD until their extinction in the early 1500s AD. The Lucayans did not come into historical light until Christopher Columbus made his first landfall in the Ancient World on 12 October 1492. Additionally, the Ancient World of the Western Hemisphere's transformation into the New World of the Americas began with Columbus' arrival in the Lucayan Islands.

The Taino culture gradually changed over time due to the ancestral transition from big-island to small-island life. After several centuries of living in the Lucayan Islands, the indigenous immigrants became a distinct cultural group.

Before European contact, the main Lucayan Islands were settled by the Lucayan people, with an estimated 20,000 to 40,000 people living in the islands.[329] *It is believed that the Lucayan population may have been as high as 80,000 people.*[330]

Over the centuries, the Lucayans adapted to their small-island limitations and hardships, including poor soil, scarce resources, and hurricanes in a warm, sub-tropical climate.[331] Nevertheless, their limited resources were supplemented by their creativity, industry, and inter-island trade.[332]

The Lucayans spoke the same language as the Taínos of Cuba, Puerto Rico, and Jamaica. However, there were differences in political structures, warfare, farming, pottery-making techniques, and natural environments.[333] Remnants of former Lucayan sites and artifacts have been found throughout The Bahamas[334] and various parts of the Turks and Caicos Islands.

Columbus described the Lucayans as "fluent in speech and intelligent."[335] They had long, straight black hair "as coarse as a

Photo 11: Replica of a Lucayan house at the Leon Levy Native Plant Preserve, Governor's Harbour, Eleuthera Island (Source: Tellis Bethel).

horse's tail" and covering their flattened foreheads as far as the eyebrows.[336] Their foreheads were flattened shortly after birth, which they believed provided added protection against blows from their enemies' weapons.[337] A child's head was placed between two flat boards, and pressure was gently applied over a sustained period until the forehead was flattened.

According to Columbus, these Native Americans were brown-skinned, physically fit in appearance, and scantily clad. The men wore loincloths, and the women had short mantle skirts.[338] They adorned themselves with body ornaments that included shells, stone beads, and bone and stone pendants. They also painted their bodies in red, black, or white colors.[339]

The islands' proximity within the Lucayan Archipelago and their closeness to the Greater Antilles (Hispaniola, Cuba, and Jamaica) made trade possible within and outside the Lucayan chain.[340] Caciques (or community chieftains) organized long-distance trade between islands where domestic goods, such as salt and dried conch, were traded using dugout canoes that transported up to 150 rowers.[341]

Scholars suggest that the Lucayans in the southern Lucayan Islands (Turks and Caicos Islands) might have had political and trade ties with the Taínos of Hispaniola. Based on evidence found at the Coralie site on Grand Turk, archaeologists theorize that tortoises, iguanas, and fishes were possibly prepared for local consumption and export to Hispaniola.[342] Cotton was valued for its practical use in trade and making hammocks. It is also believed that the Lucayans further north in the Lucayan chain had cultural and economic connections with the Taínos of Cuba.[343]

The Lucayans lived among numerous small islands and were maritime people whose villages were near the coast.[344] Unlike the larger Taíno communities centered around plazas, Lucayan sites were often found in pairs with houses on top of dunes near the ocean.[345] Their homes were made of wood and were round with either cone or rectangular-shaped roofs. The roofs were also thatched with palm branches.[346] Some of the homes accommodated extended families consisting of up to 20

Photo 12: Replica of a Lucayan canoe at Bahamas Historical Society, Nassau (Source: Tellis Bethel)

persons.[347] Remnants of former Lucayan sites and artifacts have been found throughout The Bahamas[348] and various parts of the Turks and Caicos Islands.

Their communities were less politically stratified and were led by a chief or cacique. Lucayan Caciques were chieftains responsible for the political and religious affairs[349] of their respective villages throughout the islands.[350] Unlike the mountainous, big-island life of their Taíno ancestors, the Lucayans lived in smaller villages.[351]

The four-legged, low-lying ceremonial stools (or duhos) made of stone or wood represented the Caciques' seat of authority.[352] Spiritually, the Lucayans, like their Taíno counterparts, believed in an eternal paradise called coyaba—a place of pleasure. Columbus also wrote, "We understood that they had asked us if we had come from heaven."[353]

The natives also believed in spirits called Zemis (also spelled Cemis) that lived in trees, carved images, and relics of the dead. Devils were often in the shape of a monkey or an owl. Fortune or misfortune meant that the spirits were pleased or displeased.[354]

The Lucayans were agile hunters and very skilled at diving. Up to half their food supply was from waters rich in marine resources.[355] Their reputation for diving resulted in them being exported in large numbers as slaves by the Spanish to Cubagua in 1508. Cubagua is a small island off Venezuela's Caribbean coast where the Lucayans engaged in pearl diving.[356] The foods the Lucayans ate were either indigenous to the Lucayan Islands or imported from the northern Caribbean.

Fruits eaten included pineapple, guava, tamarind, guinep, and papaya. Vegetables consisted of corn, peas, yam, and potatoes, with cassava (or manioc) being a dietary staple.[357] Meats included conch (a sea mollusk popular in the local diet today), agouti (related to rodents, called hutia by the Lucayans), fish, and turtle.[358] The Lucayans enjoyed grouper and snapper,[359] considered choice fish among today's Bahamians and Turks and Caicos Islanders.

Indigenous pottery classified as Palmetto Ware was made from burnt shells.[360] Lucayan pottery was brittle and made of a reddish, shell-tempered paste that was unique to their style.[361] The name Palmetto was coined by Charles Hoffman, who excavated the Lucayan Site at Palmetto grove on the northwest coast of San Salvador in 1965.[362]

In family life, the Lucayans were polygamists, and men could have as many wives as they were able to provide for.[363] Ancestral descent was traced through their maternal roots.[364] Lucayans also enjoyed singing and dancing for recreation, called "arieto," and drank beer made from cassava or maize.[365] They played a ball game known as "batos." During the sport, opposing teams were required to keep a rubber-like ball in the air using their feet, knees, and hips only. No hands were allowed to touch the ball.[366]

Lucayan-Taíno technical feats included making hammocks (or hamaca in Taíno) that were later installed on early European vessels for comfort and space convenience. Another indigenous practice was drying fish and meats on top of a wooden framework placed over a fire for preservation. This precursor to the modern grill was called barbacoa, from which the word barbecue was derived.[367]

Canoes (from a Taino word) were made from the trunks of trees. However, sails were not used.[368] Columbus came across a Lucayan trader in a canoe rowing from San Salvador to Rum Cay. He learned that Lucayans traveled up to 21 miles (33) in a day in their canoes and were swift rowers.[369]

The Lucayans appeared to have enjoyed a relatively more peaceful life than their Taíno ancestors in the northern Caribbean. Unfortunately, their smaller communities, lowlands, unsuitable weapons, and lack of fighting skills rendered the Lucayans defenseless against enemy attacks. (See Chapter 15, titled Invaders of the Lucayan Islands).

After eight centuries of cultural evolution in relatively peaceful environs, the Lucayan way of life would be drastically and horrifically disrupted with the arrival of Old World Europeans.

11

Pre-Columbian Explorers—Visitors to the Ancient World

There were others from the Old World of Europe, Asia, and Africa had explored the islands and continents of the Western Hemisphere centuries before Christopher Columbus' maiden voyage to the Ancient World in 1492. These groups included Norseman Leif Erikson (also spelled Eriksson, Erikson, or Ericson), the Chinese Admiral and explorer Zheng He, Africans from Africa's West Coast, and possibly others. However, none of these earlier explorers' interactions with the Ancient World reshaped global affairs as much as Columbus's historic expedition.

Leif Erikson was a Norseman from Greenland who sailed west to the Ancient World of the Western Hemisphere. According to one theory, Erikson was on a return trip from Norway to Greenland when he was blown off course and landed in the Ancient World. He called the place where he landed Vinland (the Land of Wild Grapes) on Canada's northeast coast. Vineland is believed to be in the area of the Gulf of St. Lawrence River, an outlet near Newfoundland and Nova Scotia.[370] Another theory suggests that Erikson departed Greenland on a maritime expedition to the west.[371]

Erikson subsequently made landfall in the Ancient World in 1,000 A.D, some 500 years before Columbus did and approximately 300 years after the Lucayans' ancestors had settled the Lucayan chain. Erikson's contact with Native Americans on the Western Hemisphere's northern continent during the medieval period (or Middle Ages) was short-lived.[372] His arrival in the Ancient World and the brief occupation of the Vikings who followed (including his brother Thorvald) had no lasting impact on the course of history.[373]

Illustration 11: Leif Eriksson makes landfall in North America during Middle Ages (Source: Wikipedia)

Others believe that the Chinese arrived in the Ancient World before the Europeans. In his book, *1421: The Year China Discovered the New World*, former British Royal Navy Officer

Gavin Menzies argues that Admiral Zheng He had departed China on an expeditionary voyage from 1405 to 1433. Historical records *indicate* that Admiral He's huge expeditionary fleet explored Southeast Asia, the Indian subcontinent, Arabia, and the Horn of Africa.[374]

According to Menzies, Admiral He later discovered Australia, New Zealand, and the Ancient World of the Western Hemisphere.[375] The expeditionary fleet sailed westward across the Atlantic Ocean from waters south of the African continent and arrived in the Ancient World (the Americas) approximately seventy years before Columbus.[376]

Menzies also claims that Admiral He's fleet had sailed through the Lucayan Islands.[377] Nevertheless, scholars, including Chinese authorities, argue that Menzies' theory lacks credible evidence.[378] Regardless, if the Chinese did arrive before Columbus, their activities in the Ancient World (the Americas) did not change the world.

Scholars also believe that Africans were another group of explorers who sailed to the Ancient World during the pre-Columbian era. Historical data and archaeological evidence suggest that Africans from the west coast of Africa sailed to the Ancient World as early as 200 B.C. [379]

Historians posit that Africans sailed to the Ancient World during the period Mayans were settling Central America, and the Taínos' ancestors (the Arawaks) were colonizing the Caribbean. It is further believed that the ocean currents and winds between Africa and the Ancient World made this voyage possible, as they did centuries later for Columbus and his vessels.

Based on historical accounts, the African explorers settled in what is now called South and Central America. Historians believe that Africans were present in the Ancient World during European explorations.[380] However, African immigration to the Ancient World centuries ago did not reshape global events in any permanent way.

While European Norsemen, Chinese, African, and other foreign explorers might have encountered the Ancient World before 1492, their contact with the Ancient World left the indigenous peoples, cultures, and kingdoms more or less the same way they found them.

On the other hand, it is widely accepted that a permanent global shift in human affairs occurred due to Columbus' accidental arrival in the Ancient World in 1492. Upon returning from his first voyage with the news of gold, Columbus convinced the Spanish Crown to establish a colony on Hispaniola. During his second voyage to the Americas, Columbus transported 1,200 Spanish settlers to Hispaniola in 1493.[381]

The major undertaking established the permanent presence of Europeans in the Americas, with an average of 3,000 Spanish colonists arriving annually during the early years of Spanish colonization of the Americas.[382] The sizable European exodus forever changed the Ancient World.

The organizers of the 1992 Smithsonian Seeds of Change exhibit described the unexpected encounter Columbus had with the Ancient World as "not a story of the discovery of the New World by the Old World" but rather a "story of an encounter between two branches of humankind that diverged from each other ...."[383]

This fortuitous meeting resulted in the Old World's (Europe, Africa, and Asia) reunion with the Ancient World. The events that followed also brought about the transformation of the Ancient World into the New World of the Americas by the Old World, forever changing world history.

This complete change in hemispheric affairs made the Lucayan Archipelago the site where the Americas were founded.

Historical Highlight

Greenland—The World's Largest Island

Greenland is the world's largest island and is geographically part of the Americas' northern continent. The island is south of the Arctic Ocean and east of northern Canada. Greenland was first inhabited by peoples of the Paleo-Inuit cultures, whose ancestors migrated from Alaska through Canada about 2500 years BC.

The huge island is the first territory in the Western Hemisphere permanently settled by Europeans. Europeans from Iceland and Norway began settling on the island around 986 AD, just over a decade before Leif Erickson made landfall in the North American continent. The island was ceded to Denmark in 1814 after Norway relinquished its sovereignty over it. Greenland was constitutionally administered by Denmark in 1953, making the islands' inhabitants citizens of Greenland.

Today, Greenland is an autonomous territory within the Kingdom of Denmark. The Inuit people, whose ancestors migrated from Alaska, make up the majority of the population. However, their culture is predominantly European.

Unlike Columbus' encounter, Icelandic and Norwegian arrival in the Western Hemisphere (Greenland) began during the Middle

Ages. European presence on the island did not change life in the Ancient World in any significant way.

On the other hand, Columbus' voyage to this part of the world began at the beginning of the modern era, resulting in the dismantling of an Ancient World and the birth of a new one.[384]

12

Race to the Indies

The 15th century ushered in a new age of exploration for European powers. Europeans previously paid little attention to exploring ocean waters to the west (the Ocean Sea, or today's Atlantic Ocean). Mariners were primarily set on reaching the East Indies by sailing east from South Africa's Cape of Good Hope.

Portugal, Spain's neighbor on the Atlantic coast of the Iberian Peninsula, was already occupied with maritime expeditions along the West African coast during the 1400s. Portuguese navigators intended to reach the East by sailing eastward across the Indian Ocean from South Africa. To access the wealth of the East Indies by sailing west across the Ocean Sea was a novel idea for the Old World of Columbus' generation. East Indies broadly refers to India, East Asia, and Southeast Asia, including their offshore islands.[385]

Columbus' dream took approximately seven years of planning and knocking on European monarchs' doors before he was granted sponsorship for his first transatlantic voyage. Nevertheless, his proposition was timely.

The marriage of King Ferdinand to Queen Isabella in 1469 had essentially united the Kingdoms of Castile (under Queen Isabella I) and Aragon (under King Ferdinand II),[386] which later became

Photo 13: Statue of Portuguese explorer Vasco da Gama, the first European to reach India by sea (Source: Wikipedia)

known as the Kingdom of Spain. In 1492, their united effort successfully defeated the Kingdom of Granada on the southeastern side of the Peninsula. Granada was the last of the Moorish strongholds on the Peninsula.[387] The Iberian Moors were Muslims and descendants of people from North Africa who conquered the Peninsula in 711.[388] Spain's military victory allowed the Iberian Christian kingdoms to regain complete control of the Peninsula after 700 years of Muslim occupation.[389] The conflict's ending gave Spain room to flex its influence beyond its borders.

1492 also marked the beginning of almost 400 years of Spanish exploration, conquests, and colonization of the Ancient World.[390] These conquests originated with the "discovery" of the Lucayan Islands in October of that year.

For Europe, 1492 was the closing chapter of the Middle Ages (or the medieval period, ending around 1500 AD) and the beginning of the Modern Age.[391] In some instances, Spain considers 1492 the end of the Middle Ages and the beginning of the Modern Age.[392]

Several factors drove Spain towards maritime exploration and kingdom expansion. Among these were the heated competition between European states to enlarge their kingdoms, the advancement of technology in navigation and ship design during the Renaissance era, and the monopolization of well-established overland trade routes to the Indies by Venice and the Ottoman Turks.[393]

As an emerging maritime superpower, Spain desperately wanted to beat its competition by tapping into the East Indies' gold, silk, and spice trade. Furthermore, Spain desperately needed to replenish its treasury. Columbus proposed that sailing west was an effective and efficient means of accomplishing these objectives.[394]

Portugal was another maritime superpower. The Portuguese were well underway in looking for an eastern sea route to India[395] by traveling south along the West African coast. However, fifty years of exploration by the Portuguese had only taken them as far south as the equator by the time Columbus' expedition was set in motion.[396]

Tales of gold, gemstones, silk, and spices from the Spice Islands (the Moluccas in Indonesia) by the Crusaders (1096 - 1291) and the Italian merchant and explorer Marco Polo (1271 - 1295) had piqued the interest of many Europeans in traveling to the Indies. After the fall of Constantinople in 1453, overland routes fell under the Muslim Turks' control, with high tolls for traders. Europeans then set out to discover a sea route to the East. Constantinople was the former capital of the Eastern Roman Empire in today's Istanbul, Turkey.

Spices from the Indies not only flavored European foods but also preserved them. These seasonings consisted of pepper, cinnamon, ginger, and cloves—the Arabs, who monopolized the trade, networked with merchants from Venice on Italy's northeast coast. Buying from Arab intermediaries made the goods too costly for the European market. Direct access to the Indies was more profitable for European merchants.

The most desirable spices came from the region of Indonesia. This area comprised the Malay Archipelago, consisting of thousands of islands. The Maluku Islands (or the Moluccas) are among these islands, south of the Philippines and north of Australia. Europeans called the islands the Spice Islands.[397]

During the Middle Ages, the Arabs cornered the spice markets. They controlled the nutmeg and clove markets in the Spice Islands, ginger in China, and cinnamon in India. Arab traders also established sea routes to India, Ceylon (now Sri Lanka), and the Spice Islands. On their return voyage, the Arabs sailed south of the Arabian Peninsula to Egypt through the Red Sea, between the east coast of Africa and the west coast of Arabia.

In Egypt, the Arabs traded their commodities with the merchants of Venice. Products included spice, silk, tea, and porcelain. The Venetians then sold their goods to buyers in European markets. The merchandise's exorbitant costs created a desire among Europeans, including the Venetians, to cut the Arabs out as middlemen and find a direct maritime route to the Indies to lower costs and increase profits.[398]

More than 200 years before Columbus' voyage, two brothers from Venice (Maffeo and Niccolo Polo) traveled the Far East's overland route. The Mongols were in control of Asia and parts of Europe during this period. These Asians made the roads safe for travelers in search of trade. This route was time-consuming and challenging to traverse. In 1271, the brothers went on a second journey to the Far East.

Traveling with the Polo's on their second journey was Niccolo's son Marco at the age of 17. The round-trip trek took 24 years. The Polos spent much of their time as military advisers to the Great Kublai Khan, the Mongolian emperor, and exploring the region's mainland and neighboring islands. The Great Khan was the grandson and successor of Genghis Khan and became the first Yuan Dynasty ruler of China after conquering it in 1279.[399]

During their stay, the Polos explored many parts of Asia by land and sea. Upon Marco's return to Venice in 1295, he was arrested and imprisoned for three years for commanding a Venetian vessel during a war against Genoa.[400]

While in prison, Marco's cellmate Rustichello, a writer, helped him publish a book documenting his experiences during his travels.[401] Polo was the first European to write about China, Thailand, the Maluku Islands, and other islands and areas around mainland Asia. The book, titled *Travels of Marco Polo*, was published and circulated throughout Europe.

Illustration 12: Mosaic of Marco Polo displayed in the Palazzo Doria-Tursi, in Genoa, Italy (Source: Wikipedia)

About one hundred and fifty years later, the overland trade routes to Asia were closed after the Ottoman Empire conquered the Byzantine Empire during the 1450s. This closure increased the pressure on Europeans to find a sea route to Asia.[402]

Prince Henry the Navigator's ambition was for Portugal to exploit Africa's gold and slave markets. By the 1440s, the Portuguese were fulfilling this dream. He is considered the originator of the Age of Discovery—that period when Europeans carried out maritime explorations along the west coast of Africa, the Americas, and Asia. Prince Henry was also the catalyst behind the West African slave trade's beginnings that eventually became the Atlantic Slave Trade.

The Portuguese maritime expeditions along the West African coast included gold explorations and slave-hunting. In 1441, the Portuguese obtained gold dust from Africa for the first time.[403] Also, that same year, a slave-hunting expedition was undertaken by Nuno Tristaos and Antao Goncalves. The captured Africans were taken to Portugal.[404] One of the Africans caught was an African chief, who negotiated his release with the promise to help the Portuguese capture more Africans. The agreement played a significant role in helping to initiate the West African Slave Trade.[405]

Middle East sultanates had established the East African Slave Trade well before the West African Slave Trade was initiated.[406]

Muslim Arabs sold captured Africans in North and East Africa to buyers in the Middle East through the Sahara Desert and the Indian Ocean. Arab presence in Africa grew around the 7th century with the spread of Islam in North Africa. Consequently, the African slave market existed seven centuries before European exploration of the African continent. The East African Slave Trade began to flourish during the 17th century and lasted until 1909. Zanzibar, an island off the Central East African coast, became the East African Slave Trade center.[407] It is estimated that up to 9 million East Africans or more were enslaved during this period, and over 11 million were enslaved during the West African Slave Trade.[408]

Additionally, the Barbary Slave Trade took place on the African North Coast from the early 1500s to the 1700s.[409] This trade was active during the East African Slave and West African Slave Trade eras.

Barbary (Muslim) Pirates operating out of North Africa raided European ships and coastal areas around the Mediterranean and western European countries, capturing white Europeans they sold into slavery. These pirates also traveled as far north as Iceland.

Prince Henry's dream was for Portuguese navigators to reach the east by sailing south along the West African coast, then east from South Africa to India. Prince Henry died in 1460;[410] some 28 years later, Portuguese captain Bartholomeu Dias (Diaz) rounded Africa's southern coast in 1488 but terminated his voyage before reaching India.[411] Nonetheless, Dias became the first European to reach the Indian Ocean through the Cape of Good Hope off South Africa's coast.[412]

With competition brought on by the Portuguese, who were already seeking an alternate route to Asia by sailing east from South Africa, Christopher Columbus was more than enthusiastic about fulfilling his dream of reaching the Orient by sailing west.

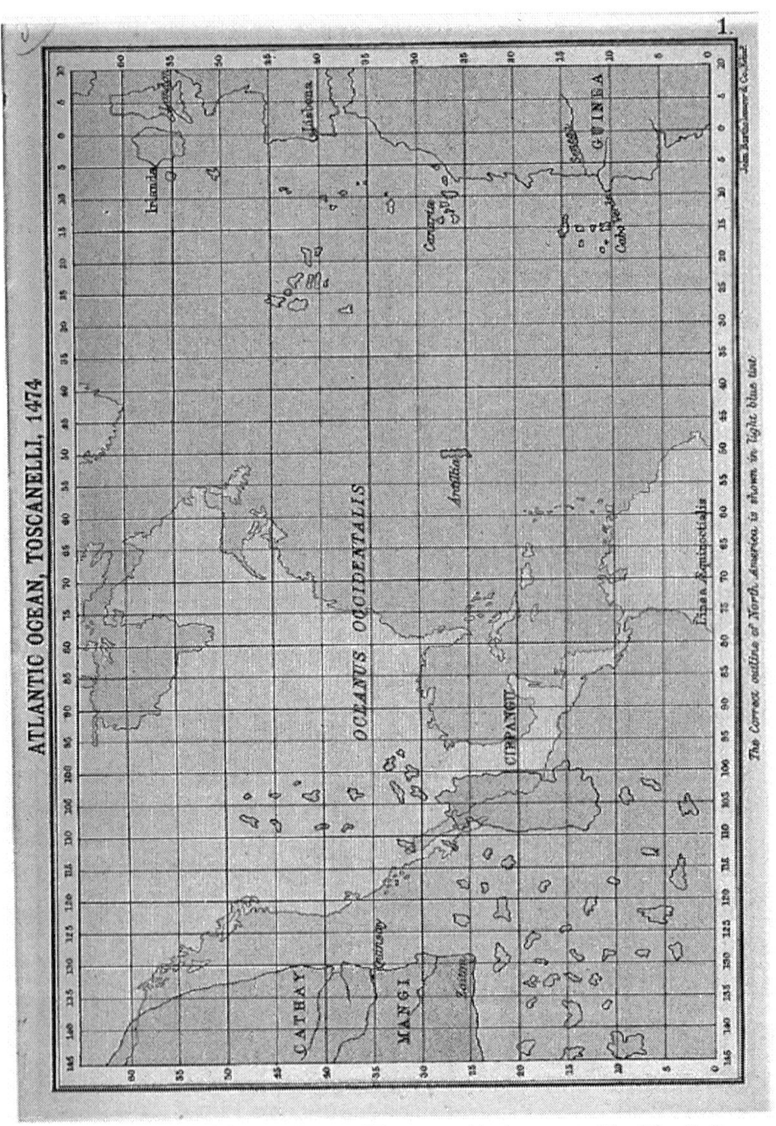

Illustration 14: Paolo dal Pozzo Toscanelli's geographical concept of the Atlantic Ocean superimposed on a modern map featuring the mythical Antilia Island in the Atlantic Ocean. Antilia is northeast of the northern Caribbean Islands. (Source: Wikimedia)

13

Admiral of the Ocean Sea

Columbus was born in the Republic of Genoa on the Italian Peninsula in 1451. In Italy, he is known as Cristoforo Colombo, and in Spain, as Cristóbal Colón. He gained his early sea experience as a navigator aboard Portuguese ships.

As a teenager, Columbus worked aboard a ship that pirates attacked in 1477.[413] The young sailor survived the ordeal after safely wading ashore. He had also sailed as far north as Iceland[414] and explored parts of the West African coast.[415]

In Portugal, he married Felipa Perestrello, and the couple had one son, Diego, who was born around 1480. Columbus' wife died shortly after giving birth. He then migrated to Spain. There, Columbus met Beatriz Enriquez de Arana, with whom he had a second son, Fernando, who was born out of wedlock in 1488.[416] Columbus and Beatriz were never married.

Columbus was an "out-of-the-box thinker" and a skilled navigator. Many held him in high esteem because of his expertise and achievements as a daring explorer.[417] Columbus was fascinated by Marco Polo's explorations and relished thoughts of amassing the wealth of the East.[418]

Photo 14: Statue of Christopher Columbus in front of Government House, Nassau, New Providence (Photo: Tellis A. Bethel)

As early as 1484,[419] about seven years before Spain had defeated the Moors on the Iberian Peninsula, Columbus devised a plan to reach the East by sailing west.[420] He was in his thirties when he began his pursuits. Columbus sought financing for his "enterprise to the Indies" from King John II of Portugal,[421] who rejected Columbus' proposal. Columbus also sought sponsorship from European authorities in England and France without success.[422]

According to one account during this period, Columbus learned that the King of Portugal had contrived a plan to sponsor a separate expedition to sail west after learning of Columbus' intentions. This duplicitous plot failed.[423]

In 1486, Columbus approached the Spanish monarchy, which rejected his initial proposition.[424] Later, the Spanish monarchs of newly liberated Spain, King Ferdinand of Aragon, Queen Isabella of Castile, and Spanish investors approved Columbus' plans in 1492. Columbus was granted an audience before the Spanish Crown with the assistance of the Spanish treasurer, Luis de Santángel, and, possibly, the Franciscan friar Juan Pérez of La Rábida, one of Queen Isabella's Confessors.[425]

For the Catholic monarchs, Columbus' transatlantic plan was a promising and potentially lucrative venture. Their European rival, the Kingdom of Portugal, was busily reaping West Africa's wealth in gold and slaves. Spain wanted its share of the action, and Columbus' ambitions matched their desires to catch up and overtake their rivals.

Columbus' first voyage was intended to be a commercial enterprise. He was subsequently commissioned to establish colonies and engage in trade in the Indies. By circumventing the hazardous overland Silk Road routes under Arab control and shortening the time to reach the Indies, Columbus aimed to reach the islands and the mainland on the opposite side of the Ocean Sea[426] (the Atlantic Ocean) before his competitors.

Columbus was elated by the Spanish Monarchs' acceptance of his proposal. His enthusiasm was expressed in his response to them:

> Your Highnesses, as Catholic Christians and Princes who love the holy Christian faith, and the propagation of it, and who are enemies to the sect of Mahoma [Islam] and to all idolatries and heresies, resolved to send me, Cristóbal Colon, to the said parts of India to see the said princes ... with a view that they might be converted to our holy faith ... Thus, after having turned out all the Jews from all your kingdoms and lordships ... your Highnesses gave orders to me that with a sufficient fleet I should go to the said parts of India.... I shall forget sleep, and shall work at the business of navigation, so that the service is performed.[427]

Illustration 13: Admiral of the Ocean Sea, Christopher Columbus (Source: Wikipedia)

Spanish authorities awarded Columbus the titles Admiral of the Ocean Sea and Viceroy of the Indies and governorship as part of the royal agreement. He was also promised 10 percent of all discovered wealth.[428] The following is an excerpt from their agreement:

> ... that of all and every kind of merchandise, whether pearls, precious stones, gold, silver, spices, and other objects and merchandise whatsoever, of whatever kind, name and sort, which may be bought, bartered, discovered, acquired and obtained within the limits of the said Admiralty, Your Highnesses grant from now henceforth to the said Don Cristóbal [Christopher Columbus] ... the tenth part of the whole, after deducting all the expenses which may be incurred therein.[429]

The medieval Castilian office of "Admiral of the Ocean Sea" was also reinstituted and awarded to Columbus. Under the responsibilities of that office, he was granted jurisdiction over maritime affairs.

Upon accepting Columbus' ambitious plan for the Indies, Spanish authorities appointed him as Governor over any lands he might discover. The Admiral of the Ocean Sea and Governor titles were to be passed down to his family, in perpetuity.[430] As the holder of these titles, Columbus became an official of the Spanish Crown, ensuring that the lands he discovered belonged to Spain.

Historical Highlight

The Lost City of 'Antilia'

Columbus' search for the Indies in the Ancient World was just as futile as the Portuguese's search for the elusive island of 'Antilia' (now spelled Antillia or Antilles). European navigators had searched for the mythical island of Seven Cities, supposedly somewhere in the Atlantic Ocean between Europe and Cipangu (Japan) during the Middle Ages.[431]

Legend had it that Antilia was settled by seven Bishops and their followers shortly after the last Visigothic king of Spain, King Roderick, was defeated by the Muslim Moors from North Africa in 711.[432] Antillia was first seen on a map produced by Genoese cartographer Battista Beccario, in 1435.[433]

The Portuguese settlers had fled from Muslim invaders on the Iberian Peninsula during the early 700s. In 1475, Portuguese authorities commissioned Fernào Teles to find the mysterious Antilia Island somewhere in the Western Ocean between Europe and East Asia.[434] The name Antilia was later modified and applied to the names "Greater Antilles" and "Lesser Antilles" in the Northern and Eastern Caribbean Islands, respectively.[435]

Illustration 15: Christopher Columbus and his men arrive at Guanahani (San Salvador) Island in the Lucayan Islands (Source: Wikimedia)

14

Columbus Makes Landfall in the Ancient World

Columbus' expedition to the Indies departed Palos de la Frontera's port on the southwestern coast of Spain on 3 August 1492. His small fleet consisted of the *Niña, Pinta* (Portuguese-style caravels), and the flagship, Santa Maria (a *carrack).*[436] The Admiral sailed aboard the flagship *Santa Maria*. He was about 41 years old when his maiden voyage began.

Nine days later, and approximately 700 miles (1126 km) southwest of Palos, Spain, Columbus arrived in the Canary Islands. The Canary Islands were a Spanish colony off Africa's northwest coast, where Columbus replenished supplies and repaired his vessel. On 5 September 1492, Columbus resumed his voyage. According to his diary, Columbus' men "lost all patience" 66 days into the journey from Spain. He encouraged the men to stay the course, reassuring them of his commitment to reaching the Indies.[437]

The next day (11 October), crewmembers saw floating debris, indicating the possibility of land with human life nearby. Within two days of crewmembers expressed frustration, a sailor named Rodrigo de Triana (also known as Juan Rodriguez Bermejo) spotted San Salvador Island aboard the *Pinta*.[438]

The island was sighted at 2 a.m. on 12 October 1492, almost 70 days after Columbus departed Spain. With over seven years in the making, Columbus' dreams of wealth, fame, and fortune were fast becoming a reality. He had outwitted his competitors and won the race to the Indies—so he thought.

Later that day, Columbus made landfall on a Lucayan Island in the Ancient World on 12 October 1492, where the indigenous Lucayans warmly received him and his entourage. He immediately took possession of the Lucayan Islands on behalf of the Spanish Crown and renamed the island San Salvador (meaning Holy Savior). The original indigenous Lucayan name for the island was Guanahani, meaning Iguana[439] or Place of the Lizard.[440]

Upon sighting land, Columbus assumed he had arrived in the Indies. On completing his first voyage in 1493, Columbus wrote to the Spanish monarchs, describing the islands he encountered as the Islands of India.[441]

Columbus immediately inquired where to find gold, having seen some of the Lucayans wearing nose rings, which he thought were made from this precious metal. It is believed that the ornaments that Columbus saw were processed copper from Hispaniola, which the Lucayans may have valued more than gold.

Supporting this belief was the name of a Taíno cacique (chief) in Hispaniola. Cacique Teréigua Hobin headed the kingdom of Xaraguá. His name meant *Prince Resplendent as Copper.*[442] Columbus noted that the Lucayans in San Salvador thought the little "gold" they had was plentiful.[443]

In response to his inquiry about the source of gold, the Lucayans pointed southward toward Cuba,[444] which Columbus presumed to be the direction of Cipangu (Japan), that place of "extensive trade, gold, spices, great shipping, and merchants."[445] Columbus crewmembers then weighed anchor and proceeded south to look for gold.

Before departing, Columbus lured several Lucayans aboard his ships as captives under the pretense that they would be repatriated to coyaba (the Lucayan word for heaven).[446] In explaining his reason for doing so, Columbus reported: "that they might learn from us, and at the same time tell us what they knew about affairs in these regions."[447]

Columbus spent about 14 days exploring the Lucayan Islands. He renamed several Lucayan Islands as he headed south of San Salvador Island.

The first island to be renamed was Wanahani (now spelled Guanahani). This island was renamed San Salvador Island (and later Watling's Island before being changed back to San Salvador in 1926).[448]

Another Lucayan Island sighted by Columbus was Haomate (today's Crooked Island), renamed La Isabella, after Queen Isabella. Additionally, Yuma (today's Long Island) was renamed Fernandina after King Ferdinand.

The remaining islands were Maniwa (today's Rum Cay), renamed Santa Maria de la Concepcion after the Virgin Mary,[449] and Hutiyakaya (today's Ragged Island), renamed Islas Arenas, meaning Island Sands.[450]

The Ragged Island chain was perhaps the last of the Lucayan Islands that Columbus saw on 27 October 1492.[451] He eventually sailed across the Great Bahama Bank's southeastern portion that now carries the name Columbus Bank.

It is unknown to what extent today's Western world was influenced by Columbus' initial encounter with the Lucayans in 1492. It is known that Columbus remained in the Lucayan Islands for two weeks and sailed with six Lucayans aboard his ships.[452] It is unknown if the abducted Lucayans remained in Hispaniola after Columbus departed the Lucayan Islands, taken to Spain on his return voyage, or made their way back to the Lucayan Islands.

Columbus would have subsequently seen or experienced various aspects of Lucayan culture—lifestyle, floras, and faunas—that would have made an indelible impression on him and his ships' companies. After all, first impressions are lasting.

Consequently, the Lucayans would have been the first people in the Ancient World to influence Western culture to some extent in language, food, manufacturing, and possibly other areas. There is no telling how many indigenous words were incorporated into the Spanish vocabulary and subsequently entered the English language due to Columbus' contact with the Lucayans.

Some indigenous words or their derivatives currently in use in today's English vocabulary, and their meanings, as follows:

Guincho - Falcon (Guinchos Cay is a name of a Bahamian cay in the southern Bahamas); Ayiti - Haiti; Sabana - Savanna;[453] Lucairi - Island People; Cayo - Cay; Caico - Outer or far away island; Cacique - Chief; Caribe - Strong or brave person; Canoa - Canoe; Huracan - Hurricane; Hamaca - Hammock; Barbacoa - Barbecue.[454] Baracutey is the name for Barracuda and describes the concept of being solitary, which is typical of this territorial fish,[455] and Higuana - Iguana.

Other cultural influences or spinoffs from the Lucayan experience would have been the canoe and the cotton hammock. The Spanish first saw hammocks on Long Island in The Bahamas.[456] Additionally, the Lucayans might have been the first to expose the Spanish to native fruits such as guava and papaya, which have Taíno names.

Nevertheless, with no tangible riches in the Lucayan Islands, Columbus departed the islands for the Northern Caribbean, never returning. He made three more voyages to the Americas over the next 11 years.

Columbus sailed further south after transiting the Columbus Bank and crossed the Old Bahama Channel. This deep-water channel separates the Great Bahama Bank's shallow waters from Cuba's northern coast. Columbus later arrived off Cuba's north coast—approximately 110 miles (178 km) south of Ragged Island—on 28 October 1492.

The date 12 October 1492 subsequently signaled the end of the Ancient World of the Western Hemisphere and the beginning of the New World of the Americas. Within a decade of Columbus' first encounter with the Ancient World, the Spanish launched slave raids in the Lucayan Islands to satisfy their quest for gold and glory.[457]

The Spanish then abandoned the islands within 30 years of Columbus' arrival after capturing nearly all of the Lucayans. England claimed the islands over a hundred years later in 1629[458] but did not settle them until 1648.[459]

15

Invaders of the Lucayan Islands

Early explorers and colonists described Lucayans as being friendly people and "lovers of peace."[460] They received Columbus with open arms after his vessels entered their waters in the Ancient World. Historical accounts of Lucayans organizing military resistance to their European captors in the Lucayan Islands are unknown.

> ... Brought [Columbus] and his landing party parrots ... and balls of cotton and spears and many other things exchanged, which they exchanged for glass beads and hawks' bells. They willingly traded everything they own... They were well-built, with good bodies and handsome features... They do not bear arms, and do not know them, for I showed them a sword, they took it by the edge and cut themselves out of ignorance. They have no iron. Their spears are made of canes.... They would make fine servants....[461]

Upon meeting the natives, Columbus learned that people from neighboring islands attacked the Lucayans. Columbus noted his observations in his diary:

I saw some who had marks of wounds on their bodies and I made signs to them asking what they were, and they showed me how people from other islands nearby came there and tried to take them, and how they defended themselves, and I believed and believe that they come from Tierra Firme (or the mainland) to take them captive.[462]

Illustration 16: Carib fishermen, illustration from Girolamo Benzoni's La historia del Mondo Nuovo (1565; History of the New World). (Source: U.S. Library of Congress)

The Lucayans reported that their ferocious enemies came from a neighboring island and ate human flesh.[463] Believing he had arrived at the East Indies' outlying islands, Columbus thought the

Lucayan enemies belonged to the Great Khan on the East Indies' mainland, where he had hoped to find vast riches.

Columbus also observed that the Lucayans were unskilled fighters with few implements of war.

"With 50 men they can all be subjugated and made to do what is required of them,"[464] Columbus reported. His comments were the first European notion of the potential enslavement of indigenous people in the Ancient World.

The Lucayans were armed with wooden spears made with fish bones and teeth at their tips, primarily fishing and hunting. Unlike the Caribs or Taínos, the Lucayans were a small ethnic group, not culturally adapted to warfare. Their inability to adequately defend themselves was possibly due to their lack of exposure to conflict. Attacks against the Tainos by the Caribs were mainly in the eastern and northern Caribbean Islands. The Lucayan Islands are geographically separated from the Caribbean and are in the northernmost reaches of the region.

The Lucayans might have become accustomed to living a relatively peaceful life, negating the need to improve their fighting skills and modernize their weapons. Nevertheless, an enemy might have been encroaching on their islands' doorstep at the time of European contact if Lucayan reports to Columbus concerning attacks by people from neighboring islands were true.

The Lucayans were easy prey with no known standing armies or warriors like the Caribs, Taínos, or Spaniards. By the time Europeans arrived, it was too late for the Lucayans to train themselves, produce better weapons, and develop a war-faring culture. Adding to the Lucayan vulnerability in defending themselves against external invasions were the tiny Lucayan Islands, flatlands and low-lying shrubs provided little protection.[465]

On the other hand, European conflicts with the Taínos did not go uncontested. The Taínos resisted European attacks with indigenous armies numbering up to 15,000 or more warriors.[466] So strong was Taino resistance to Spanish invasion that the Taino warrior Enriquillo forced the Spanish into a peace treaty, allowing the Tainos to settle in the mountains of the Dominican Republic freely. The Enriquillo-led revolt had lasted twelve years. [467]

Some researchers argue that the people who attacked the Lucayans might have been Caribs[468] or Taínos from Hispaniola.[469] Others say that Lucayan enemies might have been Native Americans from Florida.[470]

In an earlier book, *The Caribbean before Columbus*, William Keegan, curator of Caribbean archaeology at the Florida Museum of Natural History, posits that the scars were possibly due to fighting between indigenous Lucayans during inter-island trade.[471] In a later study, Ann Ross, a biological anthropologist and a professor of biological sciences at North Carolina State University, and William Keegan offer another possibility—a Carib invasion of the Lucayan Islands from Hispaniola around 1000 AD.[472]

Ross' and Keegan's theory proposes that the Carib invasion had formed the third wave of migration into the Caribbean out of northwest Venezuela or Guiana.[473] (The two previous waves were out of northeast Venezuela and the Yucatan in Central America). Ross' and Keegan's theory is based on cranial relationships between indigenous skulls found among several Caribbean Islands and mainland America. The investigation was concluded in January 2020.

Consequently, Ross' and Keegan's findings indicate that in addition to the two widely accepted routes (the Yucatan-Cuban/Belize route and the South American-Lesser Antilles route), the Caribs of the South American mainland created a third route into the Caribbean.[474]

According to Ross' and Keegan's study, mainland Caribs had canoed northward across the Caribbean Sea from the north coast of South America (Venezuela or Guiana) to Hispaniola in canoes rather than island-hop their way through the Caribbean chain. From Hispaniola, the Caribs expanded their presence into Jamaica and later the Lucayan Islands.[475] Keegan further argues that the Caribs were perhaps colonizing parts of the Lucayan Islands by the time Columbus made landfall on San Salvador.[476] Results of the research, at the time, supported their theory of the Caribs being the warlike people who invaded the Lucayan Islands.[477]

More recent genome analyses of human remains by David Reich of Harvard Medical School in collaboration with Keegan in December 2020 suggest only two migratory waves into the Caribbean from the American continent. This recent theory is derived from studies of genomes of 263 people who lived in the Caribbean and Venezuela 400 to 3,100 years ago.[478]

The research also shows that the ancient genomes were more related to those indigenous peoples who lived in Central and South America than North America. Additionally, the study indicates that the first migratory wave from mainland America arrived in Cuba about 6,000 years ago from the Central American region during the Archaic Age.

Furthermore, the analyses reveal that Arawak-speaking people from the northeastern part of South America came to Puerto Rico from around 2,500 to 3,000 years ago. This group was part of the second wave of immigrants into the Caribbean from the American mainland during the Ceramic Age. These immigrants later continued westward as they colonized the other Caribbean

Photo 15: Monument presented to The Bahamas "by the Government of Spain on the occasion of the visit of the replicas of Christopher Columbus' caravels on 10th February 1992 in commemoration of the first landfall at San Salvador in 1492." (Source: Tellis Bethel)

Islands. It is believed that the two separate groups, the Central and South American migratory groups, rarely mixed. [479]Additionally, the genome analyses suggest that the indigenous peoples within the Caribbean region were more biologically interconnected during the Ceramic Age 2,500 years ago. There were also distinct genetic differences between the Ceramic Age and Archaic Age (6,000 years ago) peoples.[480]

An analysis of Puerto Rico's and Hispaniola's populations was also carried out. Harald Ringbauer, a coauthor of the research in Reich's Harvard Lab, had developed a new technique for estimating population sizes. The test results suggest that the indigenous population within Puerto Rico and Hispaniola at the time of European contact was far less than the million or more persons initially thought to have inhabited these islands.[481]

According to the new test method, the total population count for the two Caribbean Islands is estimated at 10,000 to 50,000 indigenous people. [482] DNA analyses also indicate that all pottery styles found in the Caribbean were developed by the original settlers and their descendants and not by any new group.[483]

However, the answer to the question concerning the people who invaded the Lucayan Islands remains a mystery. Were they Native Americans from Florida, Lucayans from neighboring Lucayan Islands, Caribs, or Tainos from the Greater Antilles? Further research is needed to answer this question.

16

San Salvador Landfall—Which One?

As the island of Columbus' first landfall in the Ancient World, San Salvador would have been the first point of reference to be placed on a map representing the Ancient World. The awareness of the Lucayan Islands' existence eventually led to the "discovery" of the North and South American continents, completely reconfiguring the world map as it was known at the time.

Juan de la Cosa's map of the Americas was the first and only map produced by a cartographer with a first-hand account of San Salvador Island's geographical location. Juan de la Cosa was the owner of Columbus' vessel, the *Santa Maria,* on which he traveled when Columbus made his historic landfall on Guanahani (San Salvador). De la Cosa's map was produced in 1500, eight years after San Salvador was first sighted.[484]

San Salvador Island's exact geographic location is not clearly indicated among the other islands on Juan de la Cosa's map.[485] As a result, historians and scholars have debated for over 200 years which island in the Lucayan Archipelago is the actual site of Columbus' first landfall in the Ancient world.[486] Researchers have proposed at least eight or more possible sites in The Bahamas and Turks and Caicos Islands.[487]

In The Bahamas, prospective sites include Watling's Island[488] (named after a pirate, John Watling), which now carries the name San Salvador Island, Samana Cay, and Cat Island, all located in the central Bahamas.[489] Samana Cay is just northeast of Crooked Island and was first suggested as a landfall in 1882. The cay was later investigated as a possible site by the National Geographic team in the 1980s as the Quincentennial Celebration of Columbus' landfall (1992) was approaching.[490]

Cat Island, called San Salvador from the early 1700s until 1926, was one of the first suggestions to be put forward.[491] Another account says the name San Salvador was transferred from Watling's Island (today's San Salvador Island) to Cat Island during the early 1800s.[492] Antique maps as early as 1650 display the much smaller San Salvador (formerly Watling's Island) at its present-day location under its original Lucayan name, Guanahani, and the much larger Cat Island under its original Lucayan name, Guanama, to the northwest of San Salvador, as it is today.[493]

San Salvador Island (formerly Watling's Island) is located in the central Bahamas, approximately 350 miles (563 km) southeast of Miami, Florida. The island is approximately 13 miles (21 km) long and five miles (8 km) wide, with a population of about 940 people (2010). In 1803, the name San Salvador was transferred from Watling's Island to Cat Island as St. Salvador.[494]

Strangely enough, there is a small island located just off the northwest coast of Cat Island that still carries the name Little San Salvador (also known as Half Moon Cay)—a reminder of Cat Island's ties with Columbus' past. Today, Little San Salvador is a port of call in the central Bahamas for cruise ships of major cruise lines such as Holland America and Carnival.

In May 1926, The Bahamas' legislature renamed Watling's Island San Salvador, making that island The Bahamas' official site for Columbus' first landfall.[495]

Father Chrysostom Schreiner supported this name transfer. Father Schreiner was an American and the first permanent Catholic priest appointed to The Bahamas (1891 – 1925). He was shipwrecked on Watling's Island in January 1892.[496]

Other sites reputed to be Columbus' first landfall are found in the Turks and Caicos Islands.[497] Sites include Grand Turk and East Caicos.[498]

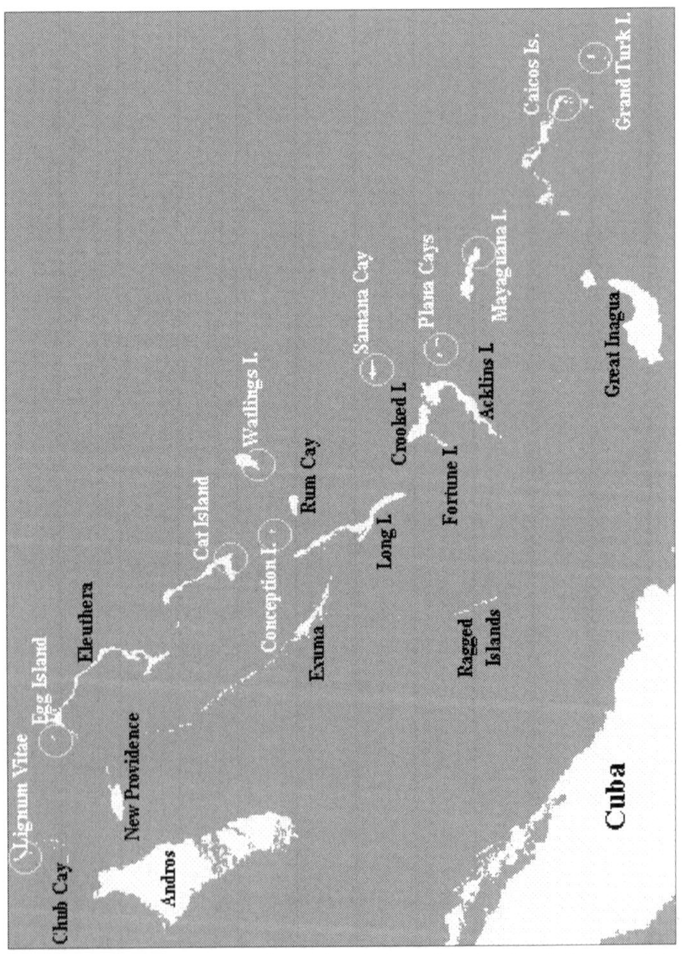

Illustration 17: Map of Lucayan Islands showing possible sites for Christopher Columbus' landfalls. (Source: Wikipedia)

Grand Turk was first suggested in 1882 by Martin Fernandez Navarrete, Spanish historian and Chief Hydrographer in Madrid, Spain. Some researchers objected to this notion based on Columbus' description of San Salvador's harbors' large capacity to accommodate many ships.[499] Others argue that there is no archaeological evidence to support the Turks and Caicos Islands' claim.[500]

To date, the archaeological excavation undertaken by Dr. Charles Hoffman from 1983 to 1985 revealed that the only site in the Lucayan chain where 15th-century Spanish paraphernalia exists alongside Lucayan items is on present-day San Salvador Island in The Bahamas.[501] The excavation took place in 1983 at the Long Bay Lucayan site, near the area where Columbus is believed to have made his first landfall. At the site, a copper blanca coin of King Henry the IV of Castile (1425 - 1474), glass beads and bronze buckles were found.[502]

Although debate continues over the location of Columbus' first landfall in the Americas,[503] most agree that the historic event took place somewhere in the Lucayan chain of islands.[504] Today, this chain of islands comprises The Bahamas and Turks and Caicos Islands.

The exact location of San Salvador on Juan de la Cosa's world map is unclear. Columbus reported that "The Island is quite big, very flat, and with very green trees and much waters and a huge lake in the middle and without any mountains." As expected, this general description has resulted in heated debates by scholars and navigators concerning which island is Columbus' landfall site. Possible sites and their proponents are in the table following:[505]

Name of Contender	Profession	Proposed Landfall
Doug Pect	Retired Air Force officer	35 miles northeast of Watling's Island
Washington Irving	Columbus biographer	Cat Island
A.B. Becher	British naval captain	Watling's Island
Gustavus Fox	Abraham Lincoln's Secretary of the Navy	Samana Cay
Joseph Judge (1986)	Associate editor of the National Geographic	Samana Cay 65 miles southeast of Watling's Island
George Gibbs	Resident of Turks and Caicos Islands	Grand Turk
Admiral Samuel Eliot Morison (1941)	Naval historian and Pulitzer Prize-winning biographer of Columbus	Guanahani
Arne Molander	Maryland historian	Egg Island, North Eleuthera
Robert Power	Californian historian	Grand Turk

Illustration 18: Christopher Columbus arrives in the Ancient World of the Western Hemisphere aboard the Santa Maria (Source: Wikipedia).

Historical Highlight

The Advent of Modern Sailing in the Ancient World

Columbus' 69-day maritime expedition comprised 85 crewmembers aboard three Spanish ships. Two of his vessels were caravels (the Niña and the Pinta), and the third vessel (the Santa Maria) was a nao.

The Santa Maria (originally named Marigalante) was owned by the Spanish navigator and cartographer of Europe's first world map, Juan de la Cosa. De la Cosa had accompanied Columbus on his first voyage to the Ancient World.[506] As a nao-type vessel, the Santa Maria was much slower than the caravels but had a larger cargo carrying capacity of 110 tons maximum.

The Santa Maria was also heavier. As Columbus' flagship, it carried 40 crewmembers.[507] The bulky vessel was about 117 feet (36 meters) long and had three masts, a forecastle, and a sterncastle. The main armament aboard the Santa Maria were bombards (a type of miniature canons) with cannonballs made of granite.[508]

The Niña and Pinta were smaller but faster than the Santa Maria. These caravels were highly maneuverable in shallow waters with a seven-foot draught. Both vessels were approximately 70 feet (21 meters) long and 21 feet (6.4 meters) wide and were crewed by 20 and 26 seamen. Their maximum cargo capacity was 50 tons. [509] The Niña and Pinta were captained by Spanish brothers, Martin Alonso Pinzón and Vicente Yañez Pinzón.

The Santa Maria ran aground and sank off Haiti's northern coast near Cap Haïtien on Christmas Eve. Wood from the wreck was used to construct Europe's first Fort in the "New World" in Hispaniola. The Pinta and Niña's whereabouts since Columbus' first voyage to the Ancient World remain unknown.[510]

The Lucayans were the first to see modern sailing vessels in the Ancient World at the beginning of a new era making the Lucayan waters the gateway to modern sailing in the Americas.

17

Columbus Ends First Voyage with Reports of Gold

Columbus departed the Lucayan Islands on 27 October 1492. The following day (28 October), he arrived off Cuba's northern coast, mistakenly identifying as Cathay (China).[511] Columbus then renamed the island "Juana" in honor of Prince Don Juan, son of Queen Isabella.[512] Columbus sailed east along Cuba's north coast.

Cuba's name is of Taíno origin and is said to mean "a center or central place."[513] The Taínos invaded Cuba from Hispaniola and occupied the island as early as 1200 AD.[514] They colonized Cuba's eastern and central areas, where they displaced the island's original inhabitants, the Guanahatebey and the Ciboney peoples (considered to be of the same ethnic group). [515]

The Guanahatebeys are thought to be among Cuba's first inhabitants from as early as 4,000 years before European arrival. These indigenous people were hunter-gatherers living in the western part of Cuba.[516] The ethnic groups (Guanahatebey/Ciboney Peoples) might have already disappeared before European explorers arrived. [517]

Columbus claimed Cuba for Spain and made his first landfall in Bariay Bay in the Province of Holguin, nearer the eastern end of Cuba's northern coast.[518] The Spanish encountered the Taínos in Gibara, west of Bariay Bay. Both Bariay and Gibara are near

Illustration 19: Columbus reception by King Ferdinand and Queen Isabella of Spain after his first return from America (Source: U.S. Library of Congress)

the east end of Cuba's northern shores.

The name Tobacco originated from among the Taíno people.[519] In Gibara, the Taínos smoked a plant-based substance that gave them the energy to work. The Taínos called the plant cohiba. The leaves were rolled and smoked like cigars. The Y-shaped pipe inserted into their nostrils while smoking was called "tabaco." The Spanish adopted the habit, which spread worldwide in the years that followed.[520] Today, Cuba is a leading producer and exporter of tobacco in the Americas.

Columbus later sailed toward the coast of Baracoa at the southeastern end of Cuba. After failing to find gold in Cuba, Columbus departed Cuba for Hispaniola (Haiti and the

Dominican Republic) on 5 December 1492. He arrived off the northwest coast of Hispaniola, which Columbus believed to be Cipangu (Japan).

The Taínos called the entire island Ayiti (Haiti), meaning Land of High Mountains.[521] Columbus renamed the island La Isla Española, meaning the Spanish Island. While at Hispaniola, Columbus' hope of finding gold was rekindled. Here, in the northern Caribbean, the Taínos presented Columbus with a decorative ornament that contained precious metal.

> Guacanagari, the Taíno chief of the Kingdom of Marien in the northern province of Hispaniola, gave Columbus "two pieces of worked gold ... very thin."[522] (The Marien territory included the northwest coast of Haiti, nearest the southern Lucayan Islands).

The gold Columbus received was nowhere near the quantity he expected to find in the New World.[523] Columbus reported, "... I believe that they get very little of it here, although I hold that they are very close to its source, and there is a great deal of it."[524]

On Christmas Eve (24 December 1492), Columbus' flagship, Santa Maria, ran aground near Haiti's northeast coast.[525] The shipwreck happened off the coast of the Marien domain. It was the first European grounding in the Americas.

Confident, nonetheless, that there was more gold to be found in the Antilles, Columbus decided to build a fort and leave 39 of his men behind. The fort was named La Navidad and constructed using material from the *Santa Maria*. La Navidad was the Europeans' first attempt to establish a permanent settlement in

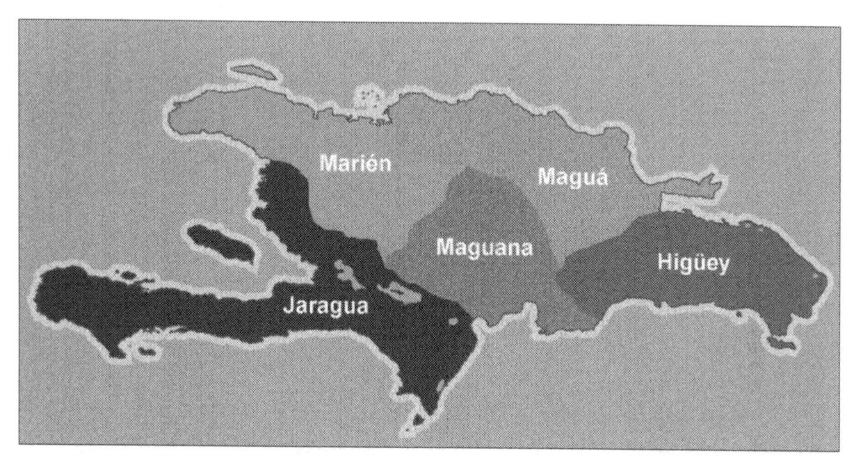

Illustration 20: Map of the island of Hispaniola (today's Haiti and the Dominican Republic) showing the five Taíno kingdoms (Source: Wikipedia)

the Americas.[526] The Fort was located on the northeast coast of today's Haiti.

In underestimating the Taínos' fighting capabilities, Columbus reported: "I have ordered a tower and fortress to be constructed and a large cellar, not because I believe there is any necessity on account of [the natives]," he noted in his journal. "I am certain the people I have with me could subjugate all this island... as the population are naked and without arms and very cowardly."

Columbus tasked his men with finding and securing gold until he returned to Hispaniola. Before departing Hispaniola, Columbus' men got into an argument with the Taínos. His men killed two Taínos, marking the Americas' first casualties by Europeans. [527]

Columbus departed Hispaniola on 16 January 1493 and arrived in Spain on 15 March 1493. He took about 10 to 25 Native Americans back with him to Spain during his first voyage.[528] About eight survived. Upon arrival, Columbus presented a small quantity of gold, native plants, and birds to Spanish authorities,[529] with exaggerated reports of gold in the Indies. Columbus claimed: "Hispaniola is a miracle ... there are many wide rivers, of which the majority contain gold. There are many spices and great mines of gold and other commodities."[530]

Before arriving in Palos, Spain, Columbus stopped in Portugal. Upon learning of Columbus' discovery in the Americas, the King of Portugal accused Spain of infringing on his kingdom's domain.[531] Portugal believed it held rights to lands discovered in the Atlantic Ocean under the Treaty of Alcáçovas, signed in 1479.

Under this Treaty, Spain was granted the Canary Islands, and Portugal took possession of the Azores, Cape Verde Islands, and Madeira. As a result, Portugal considered the Lucayan Islands to be Portugal's territory and not as the Spanish Crown claimed.

The 1479 Treaty had divided the world into a north-south boundary, with Portugal being granted all lands found south "from the Canary Islands down toward Guinea" on the West African coast. The Treaty also gave Portugal rights to monopolize trade on the West African coast, including the slave trade.[532]

After the Portuguese had disputed Spain's claims in the Americas, the Spanish monarchs immediately sought the Spanish-born Pope Alexander VI's support to establish Spain's territorial claims.[533] Boundaries between Spain's and Portugal's territories were subsequently settled by a Papal Bull (an official proclamation) in 1493.[534]

Consequently, a longitudinal line was drawn 100 leagues (320 miles or 514 km) west of the Portuguese-controlled Cape Verde Islands or Cabo Verde in Portuguese. (A longitudinal line is an imaginary line drawn from the North Pole to the South Pole).

Additionally, Pope Alexander VI's Bull declared that all territories west and south of this longitudinal line were Spain's, and all regions east of this line were Portugal's. [535] Cape Verde is an archipelago in the Atlantic Ocean 350 miles (570 km) west of Senegal in northwest Africa. Spain and Portugal had divided the world's lands and seas without regard for the indigenous peoples who already occupied them or their European counterparts.

Coincidentally, there is a tiny cay on the eastern boundary of the Great Bahama Bank's southeastern border. Its name is Cay

Verde, and it stands as a reminder of a world whose territories were geographically divided between two European superpowers. Additionally, this portion of the Great Bahama Bank is named the Columbus Bank, another reminder of Columbus' presence in the Lucayan Islands.

Dissatisfied by Pope Alexander's territorial allocations, Portugal renegotiated directly with Spain for the line of demarcation to be moved an additional 270 leagues west of the Cay Verde Islands in 1494. The new negotiations extended Portugal's territory over 900 miles (1448 km) farther west of the Canary Islands.[536] Spain and Portugal concluded talks under the Treaty of Tordesillas between themselves which, unknowingly to both parties, allocated Brazil to Portugal.[537]

Columbus' "discovery" of the "Ancient World" subsequently resulted in two European superpowers—Spain and Portugal—dividing global lands and oceans amongst themselves without regard for the indigenous peoples who inhabited them.

With the Ancient World falling under Spain's jurisdiction, the Spanish Monarchs authorized its colonization and exploitation, eventually transforming it into the Americas. Regrettably, the Lucayans would be the first among many casualties to suffer genocide during the unfolding of this New World. The story of these Lucayan Islands continues in the second book in this series, *The Lucayan Sea—Birthplace of the Americas.*

Timeline: Turks and Caicos Islands (8th Century to 1976)

700s – 1520: Taínos and Lucayans occupy the Turks and Caicos Islands.

1492: Christopher Columbus arrived in the Lucayan Islands.

1500 – 1900: Turks and Caicos Islands claimed by Spain, France, and eventually Britain.

1512: Ponce de Leon sails off the coast of Turks and Caicos Islands en route to Florida.

1670s: Visited seasonally by Bermudan salt rakers.

1690s – 1780s: Bahamas Colony periodically attempted to tax Turks and Caicos Islands' salt industry.

1690 – 1720: Infested by pirates.

1706: Turks Island captured by Spanish and French troops.

1710: Turks Islands recaptured by Bermuda Colonists.

1718: Women Pirates Anne Bonny and Mary Read settle on Parrot Cay.

1764: The British Crown claims ownership of the Turks and Caicos Islands.

*1766 – 1848: Turks and Caicos Islands falls under Bahama Colony jurisdiction.

1766: First King's Agent, Andrew Symmer, appointed to Turks and Caicos Islands.

1780s: Many American colonists loyal to the British Crown move to The Bahamas and Turks and Caicos Islands due to the War of Independence between American colonists and Britain. The American loyalists also bring their African slaves with them.

1783: French troops invaded Turks Island. British Admiral Horatio Nelson was unsuccessful in defeating the French.

1783: Turks and Caicos Islands granted to Britain under the Treaty of Versailles.

1813 & 1815: Turks and Caicos Islands severely damaged by hurricanes.

1846 – 1883: Whaling industry established on Grand Turk.

1848 - 1873: The British allow Turks and Caicos Islands self-government.

1848: Turks and Caicos Islands became a Crown colony under Jamaica's governorship.

1866: The Great Bahama hurricane devastates the entire Lucayan chain.

1874 – 1962: Jamaica annexes Turks and Caicos Islands until Jamaica (a British colony) becomes an independent country.

1945: First chief minister of the Turks and Caicos Islands, James Alexander George Smith ('Jags') McCartney (1976 – 1980), was born on Grand Turk, Turks and Caicos Islands.

1948: First commercial flight to Turks and Caicos Islands.

1950s: Many Turks and Caicos Islanders moved to The Bahamas for employment in Freeport, Grand Bahama Island, which was transformed into a free trade zone.

1959: New constitution and constitutional elections introduced in the Turks and Caicos Islands. Post of administrator also re-instituted.

1959: Turks and Caicos Islands have their administration under the governorship of Jamaica.

1962: Jamaica becomes an independent country; The Bahama Colony is granted jurisdiction over the Turks and Caicos

Islands for a second time until the Bahama Colony became independent in 1973.

1962: The Turks and Caicos Islands become Crown Colony after Jamaica is granted independence.

1965: Turks and Caicos Islands fall under the jurisdiction of the Bahama Colony.

1966: Tourism takes root in the Turks and Caicos Islands, with American investors constructing airstrip and the first hotel in Providenciales, The Third Turtle.

1962: Astronaut John Glenn arrived ashore at Grand Turk after the first human-crewed space flight.

1976: The founder and leader of the Peoples' Democratic Movement, Mr. James Alexander George Smith ('JAGS') McCartney, becomes the first chief minister of the Turks and Caicos Islands until his death in a plane crash in 1980. Mr. Oswald Skippings succeeded Mr. McCartney as chief minister.

References

[1] Stausberg, Michael (2011). Religion and Tourism: Crossroads, Destinations and Encounters. New York: Routledge, p.128.
[2]https://www.ourdocuments.gov/doc.php?flash=false&doc=23 [Retrieved: 30 November 2019].
[3] https://www.nytimes.com/1964/10/03/archives/hartford-to-sell-paradise-island-ap-heir-offers-resort-in-bahamas.html [Retrieved: 30 November 2019].
[4]https://en.wikipedia.org/wiki/Paradise_Lost#cite_note-6 [Retrieved: 30 November 2019].
[5]https://www/atlantisbahamas.com/things-to-do/marine-habitat [Retrieved: 13 February, 2021]
[6] http://ssmaritime.com/oceanexplorer.htm [Retrieved: 29 November, 2019].
[7]https://www.nasa.gov/mission_pages/LRO/multimedia/lroimages/lola-20100528-maretranquillitatis.html [Retrieved: 10 June 2020].
[8] https://en.wikipedia.org/wiki/Lucayan_Archipelago [Retrieved 9 June, 2014].
[9] Smith, Jean Reeder and Smith, Baldwin Lacey. (1980). *Essentials of World History.* New York: Barron's Educational Series, Inc. p.iii. See also: Duncan, Marcel. (1964). *Larousse Encyclopedia of Modern History, From 1500 to the Present Day*. New York: Harper and Row.
[10] Smith, Jean Reeder and Smith, Lacey Baldwin (1980). Essentials of World History. London: Barron Educational Series, pp.76-77.
[11]https://www.britannica.com/event/Middle-Ages [Retrieved: 17 January 2021].
[12] Kamen, Henry. (2005). *Spain 1469–1714* (3rd Ed.). New York: Pearson/Longman. p. 29.
[13] https://www.merriam-webster.com/dictionary/eastern%20hemisphere [Retrieved: 17 January, 2021].
[14] https://www.newyorker.com/culture/culture-desk/canadas-impossible-acknowledgment [Retrieved: 17 January 2021].
[15] https://www.britishempire.co.uk/maproom/turksandcaicos.htm [Retrieved: 17 January 2021].
[16] https://www.britannica.com/topic/Commonwealth-association-of-states [Retrieved 10 June 2020].

[17] Hutchins, C.D. (1977). The Story of the Turks and Caicos Islands. Wisconsin: University of Wisconsin. p.8.
[18]https://www.nasa.gov/mission_pages/LRO/multimedia/lroimages/lola-20100528-maretranquillitatis.html [Retrieved: 10 June 2020].
[19] https://www.express.co.uk/news/science/1154975/Moon-landing-how-far-away-is-Moon-how-many-miles-did-Apollo-11-travel-NASA-news [Retrieved 10 June 2020].
[20] Sauer, C.O. (1966).*The Early Spanish Main.* California: University of California Press. pp.159-160.
[21] https://www.geraceresearchcentre.com/pdfs/14thNatHist/197-211_Gnivecki.pdf [Retrieved 4 April 2020].
[22] Parker, Christopher. (2001). *Bahamas and Turks and Caicos.* (2nd ed.). Victoria, Australia: Lonely Planet Publications, p.28.
[23] http://www.tribune242.com/news/2015/apr/28/most-beautiful-place-space/ [Retrieved 3 April 2020]
[24] https://gizmodo.com/tongue-of-the-ocean-1550931325 [Retrieved 10 June 2020].
[25]https://en.wikipedia.org/wiki/Atlantic_Undersea_Test_and_Evaluation_Center [Retrieved 10 June 2020].
[26] https://www.etymonline.com/search?q=archipelago [Retrieved 4 April 2020].
[27] https://www.bahamas.com/vendor/tropic-cancer-beach [Retrieved 10 June 2020].
[28] Thompson, G.A. (1812). The Geographical and Historical Dictionary of America and the West Indies, Volume 1. London: Harding and Wright. p.124.
[29] Keegan, William F. and Carlson, Lisabeth A. (2008). Talking Taino: Caribbean Natural History from a Native Perspective. Alabama: University of Alabama Press, p.11.
[30] Keegan, William F. and Carlson, Lisabeth A. (2008). *Talking Taino: Caribbean Natural History from A Native Perspective.* Alabama: University of Alabama Press.pp.11-12.
[31] Keegan, William F. and Carlson, Lisabeth A. (2008). *Talking Taino: Caribbean Natural History from a Native Perspective. Alabama*: University of Alabama Press, p.11.
[32] https://curiosity.lib.harvard.edu/scanned-maps/catalog/44-990127828200203941 [Retrieved: 20 April 2020].
[33] West India Islands. Caribbean Lucayas Windward/Leeward Is. Johnston 1906 map. *The World Wide Atlas of Modern Geography Political and Physical* by J. Scott Keltie. Published by W. & A. K. Johnston, Edinburgh & London. 7th Edition.

[34] *The Study of Lucayan Duhos*. Joanna Ostapkowicz, World Museum Liverpool. https://www.floridamuseum.ufl.edu/wp-content/uploads/sites/44/2017/04/JCA_ost_final_pub2.pdf [Retrieved 5 May 2020].
[35] Keegan, William F. and Carlson, Lisabeth A. (2008). *Talking Taino: Caribbean Natural History from a Native Perspective*. Alabama: University of Alabama Press, p.21.
[36] https://www.loc.gov/resource/g3932c.ar160802/ [Retrieved 10 June 2020].
[37] Riley, Sandra. (2000). Homeward Bound: A History of the Bahama Islands to 1850 with a Definitive Study of Abaco In the American Loyalist Plantation Period. Florida: Riley Hall Publishers. pp.1-2.
[38] Riley, Sandra. (2000). Homeward Bound: A History of the Bahama Islands to 1850 with a Definitive Study of Abaco In the American Loyalist Plantation Period. Florida: Riley Hall Publishers. p.1.
[39] Riley, Sandra. (2000). Homeward Bound: A History of the Bahama Islands to 1850 with a Definitive Study of Abaco In the American Loyalist Plantation Period. Florida: Riley Hall Publishers. p.1.
[40] Meyerhoff A.A. and Hatten C.W. (1974). Bahamas salient of North America: Tectonic framework, stratigraphy and petroleum potential. AAPG Bull. 58:1201–1239.
[41] https://www.ncbi.nlm.nih.gov/pmc/articles/PMC2606802/ [Retrieved 5 April 2020].
[42] https://www.caribbeanauthority.com/how-were-the-caribbean-islands-formed/ [Retrieved 20 April 2020].
[43]https://www.sam.usace.army.mil/Portals/46/docs/military/engineering/docs/WRA/
Bahamas/BAHAMAS1WRA.pdf [Retrieved 20 April 2020].
[44]https://www.geraceresearchcentre.com/pdfs/GeologyKarstSanSal_JMylroieJCarew.pdf [Retrieved 20 April 2020].
[45] https://www.aquaticadventures.com/silver-bank-dominican-republic [Retrieved 10 June 2020].
[46] Riley, Sandra. (2000). Homeward Bound: A History of the Bahama Islands to 1850 with a Definitive Study of Abaco In the American Loyalist Plantation Period. Florida: Riley Hall Publishers. p. 2.
[47] Henry, James Stark. (1891). Stark's History and Guide to the Bahama Islands Containing a Description of Everything on or About the Bahama Islands of Which the Visitor or Resident May Desire Information, Including Their History, Inhabitants, Climate, Agriculture, Geology, Government and Resources. Massachusetts: H. M. Plimpton A Co., Printers A Binders. p.130.

[48]https://www.sam.usace.army.mil/Portals/46/docs/military/engineering/docs/WRA/
Bahamas/BAHAMAS1WRA.pdf [Retrieved 20 April 2020].
[49] https://bahamasgeotourism.com/entries/the-hermitage-on-mt-alvernia/751b6ee7-a764-40bb-8e9b-5d7de8bfc5a0 [Retrieved 10 June 2020].
[50]https://www.sam.usace.army.mil/Portals/46/docs/military/engineering/docs/WRA/Bahamas/
BAHAMAS1WRA.pdf [Retrieved 20 April 2020].
[51] https://en.wikipedia.org/wiki/Pico_Duarte [Retrieved 10 June 2020].
[52] https://en.wikipedia.org/wiki/Pico_Turquino [Retrieved 10 June 2020].
[53]https://www.nationsencyclopedia.com/Americas/index.html#ixzz6HuGvFzZn [Retrieved 24 March 2020].
[54] https://www.nationsencyclopedia.com/Americas/United-Kingdom-American-Dependencies-TURKS-AND-CAICOS-ISLANDS.html#ixzz6HuIlnIJH [Retrieved 24 March 2020].
[55] https://www.nationsencyclopedia.com/Americas/Cuba-LOCATION-SIZE-AND-EXTENT.html#ixzz6HuGbfc4i [Retrieved 24 March 2020].
[56] https://www.best.gov.bs/wp-content/uploads/2016/03/Bahamas-NAR.pdf [Retrieved 10 June 2020].
[57] Riley, Sandra. (2000). Homeward Bound: A History of the Bahama Islands to 1850 with a Definitive Study of Abaco In the American Loyalist Plantation Period. Florida: Riley Hall Publishers. p. 2.
[58] https://www.britannica.com/place/Turks-and-Caicos-Islands [Retrieved 10 March 2020].
[59]https://www.sam.usace.army.mil/Portals/46/docs/military/engineering/docs/WRA/Bahamas/
BAHAMAS1WRA.pdf [Retrieved: 20 April 2020].
[60]
https://www.un.org/depts/los/nippon/unnff_programme_home/fellows_pages/fellows_papers/turnquest_0506_bahamas.pdf [Retrieved 5 May 2020].
[61] Ladd, Harry, S. Chairman. (1948). Report of the Subcommittee on the Ecology of Marine Organisms By National Research Council (U.S.). Subcommittee on the Ecology of Marine Organisms. Washington D.C.: Division of Geology and Geography, National Research Council. p. 37.
[62] Ladd, Harry, S. Chairman. (1948). Report of the Subcommittee on the Ecology of Marine Organisms By National Research Council (U.S.). Subcommittee on the Ecology of Marine Organisms. Washington D.C.: Division of Geology and Geography, National Research Council. p. 37.

[63] https://www.calcean.com/oolitic-aragonite [Retrieved 10 June 2020].
[64] https://www.bahamapundit.com/2017/01/aragonite-again.html [Retrieved: 17 January 2020].
[65] http://turksandcaicostourism.com/about-turks-and-caicos/quick-facts/ Retrieved 20 March 2020].
[66] https://www.britannica.com/place/Turks-and-Caicos-Islands [Retrieved 20 March 2020].
[67] https://www.britannica.com/place/Turks-and-Caicos-Islands [Retrieved 20 March 2020].
[68] https://suntci.com/discrediting-the-myth-tci-dependent-or-not-p5671-129.htm [Retrieved: 30 December 2020].
[69] https://www.caribjournal.com/2019/02/27/bahamas-blue-hole-capital-world/ [Retrieved 20 April 2020].
[70]https://www.sam.usace.army.mil/Portals/46/docs/military/engineering/docs/WRA/Bahamas/
BAHAMAS1WRA.pdf [Retrieved 20 April 2020].
[71] https://www.geraceresearchcentre.com/pdfs/14thNatHist/151-172_Berman.pdf [Retrieved 20 April 2020].
[72] https://www.britannica.com/place/The-Bahamas/Climate [Retrieved 26 March 2020].
[73]https://www.sam.usace.army.mil/Portals/46/docs/military/engineering/docs/WRA/Bahamas/
BAHAMAS1WRA.pdf [Retrieved 20 April 2020].
[74] https://www.britannica.com/place/Turks-and-Caicos-Islands [Retrieved 20 March 2020].
[75] https://www.britannica.com/place/Turks-and-Caicos-Islands [Retrieved 20 March 2020].
[76] https://www.britannica.com/place/Turks-and-Caicos-Islands [Retrieved 20 March 2020].
[77]https://www.sam.usace.army.mil/Portals/46/docs/military/engineering/docs/WRA/Bahamas/ BAHAMAS1WRA.pdf [Retrieved 20 April 2020].
[78] https://www.montgomerybotanical.org/event-posts/zamia-lucayana-outreach-in-the-bahamas/ [Retrieved 20 May 2020].
[79] http://animaldiversity.org/accounts/Geocapromys_ingrahami/ [Retrieved 27 August 2020].
[80] https://en.wikipedia.org/wiki/Cyclura_rileyi [Retrieved 27 August 2020].

[81] https://blueprojectatlantis.org/lucaya-is-on-the-move-tour-de-turtle-2018/ [Retrieved 20 April 2020].
[82] https://en.wikipedia.org/wiki/Lucayan_Archipelago [Retrieved 17 April 2021]. See also: Granberry, Julian, & Gary Vescelius (2004). Languages of the Pre-Columbian Antilles. Tuscaloosa, AL: University of Alabama Press.
[83] Craton, Michael and Saunders, Gail. (1999). Islanders in the Stream: A History of the Bahamas. Vol. One: From Aboriginal Times to the End of Slavery. Georgia: Georgia University Press. p. 393.
[84] Riley, Sandra. (2000). Homeward Bound: A History of the Bahama Islands to 1850 with a Definitive Study of Abaco In the American Loyalist Plantation Period. Florida: Riley Hall Publishers. pp. 26-27.
[85] Craton, Michael and Saunders, Gail. (1999). Islanders in the Stream: A History of the Bahamas. Vol. One: From Aboriginal Times to the End of Slavery. Georgia: Georgia University Press. p. 64.
[86] Riley, Sandra. (2000). Homeward Bound: A History of the Bahama Islands to 1850 with a Definitive Study of Abaco In the American Loyalist Plantation Period. Florida: Riley Hall Publishers. pp. 26-27.
[87] Saunders, Nicholas. (Ed.) (2005). The Peoples of the Caribbean: An Encyclopedia of Archaeology and Traditional Culture. ABC-CLIO, Inc.: California, p.157.
[88] https://en.wikipedia.org/wiki/Robert_Heath [Retrieved 20 April 2020].
[89] Riley, Sandra. (2000). Homeward Bound: A History of the Bahama Islands to 1850 with a Definitive Study of Abaco In the American Loyalist Plantation Period. Florida: Riley Hall Publishers. pp. 26-27.
[90] Riley, Sandra. (2000). Homeward Bound: A History of the Bahama Islands to 1850 with a Definitive Study of Abaco In the American Loyalist Plantation Period. Florida: Riley Hall Publishers. p. 28.
[91] https://www.history.com/news/pilgrims-puritans-differences [Retrieved 20 April 2020].
[92] https://www.britannica.com/topic/Puritanism [Retrieved 17 January 2021].
[93]www.bbc.co.uk/history/british/civil_war_revolution/choosingsides_01.shtml [Retrieved 17 January 2021].
[94] https://en.wikipedia.org/wiki/ [Retrieved 17 January 2021].
[95] Riley, Sandra. (2000). Homeward Bound: A History of the Bahama Islands to 1850 with a Definitive Study of Abaco In the American Loyalist Plantation Period. Florida: Riley Hall Publishers. p. 28.

[96] Craton, Michael and Saunders, Gail. (1992). Islanders in the Stream: A History of the Bahamian People-From Aboriginal Times to the End of Slavery. Georgia: University of Georgia Press, pp. 73-76.
[97] https://www.britannica.com/place/The-Bahamas [Retrieved 5 May 2020].
[98] Cash, Phillip and Maples, Don. (1987). The Making of The Bahamas: A History for Schools. England: Longman Caribbean. p.22.
[99] Cash, Phillip and Maples, Don. (1987). The Making of The Bahamas: A History for Schools. England: Longman Caribbean. p.22.
[100] Craton, Michael and Saunders, Gail. (1992). Islanders in the Stream: A History of the Bahamian People-From Aboriginal Times to the End of Slavery. Georgia: University of Georgia Press, pp. 73-76.
[101] https://www.britannica.com/place/The-Bahamas/British-colonization [Retrieved 5 May 2020].
[102] https://www.sun-sentinel.com/news/fl-xpm-1992-05-19-9202090204-story.html [Retrieved 20 April 2020].
[103] https://en.wikipedia.org/wiki/Charles_II_of_England [Retrieved 20 April 2020].
[104] https://www.britannica.com/event/English-Civil-Wars [Retrieved 20 April 202].
[105] https://www.britannica.com/biography/Oliver-Cromwell [Retrieved 20 April 2020].
[106] https://www.post-gazette.com/local/south/2010/11/24/Native-Americans-played-crucial-role-in-settlers-survival/stories/201011240253 [Retrieved 20 April 2020].
[107] Curry, Robert. (1930). Bahamian Lore. Paris: (Privately Printed). pp. 115 - 122. See: http://www.jabezcorner.com/Grand_Bahama/1647_articles.htm [Retrieved 10 May 2020]. See also: https://bahamianology.com/william-sayle-bahamas-history/ [Retrieved 20 April 2020].
[108] Riley, Sandra. (2000). Homeward Bound: A History of the Bahama Islands to 1850 with a Definitive Study of Abaco In the American Loyalist Plantation Period. Florida: Riley Hall Publishers. p. 35.
[109] Riley, Sandra. (2000). Homeward Bound: A History of the Bahama Islands to 1850 with a Definitive Study of Abaco In the American Loyalist Plantation Period. Florida: Riley Hall Publishers. pp. 32-33.
[110] https://harvardmagazine.com/2010/05/cast-your-bread [Retrieved 10 May 2020].

[111] Riley, Sandra. (2000). Homeward Bound: A History of the Bahama Islands to 1850 with a Definitive Study of Abaco In the American Loyalist Plantation Period. Florida: Riley Hall Publishers. p. 31.
[112] https://en.wikipedia.org/wiki/The_Restoration [Retrieved 20 April 2020].
[113] https://en.wikipedia.org/wiki/Charles_II_of_England [Retrieved 20 April 2020].
[114] https://www.carolana.com/Carolina/Governors/wsayle.html [Retrieved 17 January 2021].
[115] Riley, Sandra. (2000). Homeward Bound: A History of the Bahama Islands to 1850 with a Definitive Study of Abaco In the American Loyalist Plantation Period. Florida: Riley Hall Publishers. pp. 31-32.
[116] Riley, Sandra. (2000). Homeward Bound: A History of the Bahama Islands to 1850 with a Definitive Study of Abaco In the American Loyalist Plantation Period. Florida: Riley Hall Publishers. pp. 31-32.
[117] Riley, Sandra. (2000). Homeward Bound: A History of the Bahama Islands to 1850 with a Definitive Study of Abaco In the American Loyalist Plantation Period. Florida: Riley Hall Publishers. p. 33.
[118] Riley, Sandra. (2000). Homeward Bound: A History of the Bahama Islands to 1850 with a Definitive Study of Abaco In the American Loyalist Plantation Period. Florida: Riley Hall Publishers. pp. 33-34.
[119] Lefroy, Sir John, Henry. (1879). Memorials of the Discovery and Early Settlement of the Bermudas Or Somers Island (1511 – 1687), Volume 2. London: Longmans, Green & Co. p.255.
[120] Bethell, Talbot. (1937). The Early Settlers of the Bahamas and Colonists of North America. Maryland: Heritage Books, Inc. p.63.
[121] https://www.carolana.com/Carolina/Governors/wsayle.html [Retrieved 17 January 2021].
[122] https://www.britannica.com/place/The-Bahamas/British-colonization [Retrieved 1 May 2020].
[123] https://en.wikipedia.org/wiki/Charles_II_of_England [Retrieved 20 April 2020].
[124] https://en.wikipedia.org/wiki/Charles_II_of_England [Retrieved 20 April 2020].
[125] https://en.wikipedia.org/wiki/Charles_II_of_England [Retrieved 20 April 2020].
[126] https://www.newnetherlandinstitute.org/history-and-heritage/digital-exhibitions/a-tour-of-new-netherland/albany/fort-nassau/ [Retrieved 17 January 2021].
[127] https://en.wikipedia.org/wiki/Nassau_ (village)_New_York [Retrieved 17 January 2021].

[128] https://en.wikipedia.org/wiki/Nassau_ (town)_New_York [Retrieved 17 January 2021].
[129] https://en.wikipedia.org/wiki/Bernardo_de_G%C3%A1lvez
[130] Riley, Sandra. (2000). Homeward Bound: A History of the Bahama Islands to 1850 with a Definitive Study of Abaco In the American Loyalist Plantation Period. Florida: Riley Hall Publishers. p. 39.
[131] Riley, Sandra. (2000). Homeward Bound: A History of the Bahama Islands to 1850 with a Definitive Study of Abaco In the American Loyalist Plantation Period. Florida: Riley Hall Publishers. p. 39.
[132] Parry, Dan. (2006). Blackbeard: The Real Pirate of the Caribbean. North Carolina: National Maritime Museum. p.126. See: The National Geographic Society. (1936). The National Geographic Magazine, Volume 69. Washington D.C.: National Geographic Society. p. 234.
[133] Johnson, Whittington and Race, B. (2000). Relations in the Bahamas, 1784-1834: The Nonviolent Transformation from a Slave to a Free Society. Arkansas: University of Arkansas Press. p. 1.
[134] https://www.britannica.com/place/The-Bahamas/British-colonization [Retrieved 2 April 2020].
[135] Fansworth, P. (1996). The Influence of Trade on Bahamian Slave Culture. *Historical Archaeology*, 30(4), 1-23.
[136] Johnson, Whittington and Race, B. (2000). Relations in the Bahamas, 1784-1834: The Nonviolent Transformation from a Slave to a Free Society. Arkansas: University of Arkansas Press. p. 1.
[137] Craton, Michael and Saunders, Gail. (1992). Islanders in the Stream: A History of the Bahamian People: Volume Two: From Aboriginal Times to the End of Slavery. Georgia: University of Georgia Press. p. 231.
[138] https://www.mybahamasponge.shop/sponge-history/ 17 January 2021].
[139] Craton, Michael and Saunders, Gail. (1998). Islanders in The Stream: A History of the Bahamian People from the Ending of Slavery to the Twenty-First Century. Vol. II. Georgia: University of Georgia Press. p.151.
[140] https://www.tourismtoday.com/about-su/tourism-history [Retrieved 20 April 2020].
[141] Johnson, Doris, L. (1972). *The Quiet Revolution in The Bahamas.* Nassau, Bahamas: Family Islands Press. p.1 70.
[142] https://en.wikipedia.org/wiki/Politics_of_the_Bahamas [Retrieved 20 April 2020].
[143] 2010 Census: See: http://statistics.bahamas.gov.bs/download/023796600.pdf [Retrieved 20 April 2014].

[144] 2010 Census: See: https://www.paho.org/hq/dmdocuments/2012/2012-hia-turks.pdf [Retrieved 10 June 2020].
[145] https://www.worldometers.info/world-population/bahamas-population/ [Retrieved 10 June 202].
[146] https://www.best.gov.bs/wp-content/uploads/2016/03/Bahamas-NAR.pdf [Retrieved 10 June 2020].
[147] https://www.caribjournal.com/2020/01/23/bahamas-tourism-record-arrivals/ [Retrieved 1 May 2020].
[148] Mills, Carlton (Ed.). (2008). *A History of the Turks and Caicos Islands.* Oxford: MacMillan Publishers, p. 2.
[149] De Booy, T. (1918). The Turks and Caicos Islands, British West Indies. Geographical Review, 6(1), 37-51. doi:10.2307/207448.
[150] https://suntci.com/discrediting-the-myth-tci-dependent-or-not-p5671-129.htm [Retrieved 7 April 2020]
[151] https://onlinelibrary.wiley.com/doi/full/10.1111/j.1540-6563.2007.00178.x [Retrieved 7 April 2020].
[152] https://www.tcmuseum.org/culture-history/salt-industry [Retrieved 7 April 2020].
[153] Mills, Carlton (Ed.). (2008). *A History of the Turks and Caicos Islands.* Oxford: MacMillan Publishers. pp. 132, 250.
[154] https://www.smithsonianmag.com/history/white-gold-how-salt-made-and-unmade-the-turks-and-caicos-islands-161576195/ [Retrieved 7 April 2020]
[155] https://onlinelibrary.wiley.com/doi/10.1111/j.1540-6563.2007.00178.x [Retrieved 7 April 2020].
[156] Kennedy, Cynthia M. (2007). The Other White Gold: Salt, Slaves, the Turks and Caicos Islands, and British Colonialism. The Historian. Volume 69, Issue2, Summer 2007. Pages 215-230. (See: https://onlinelibrary.wiley.com/doi/full/10.1111/j.1540-6563.2007.00178.x [Retrieved 7 April 2020].
[157] Kennedy, Cynthia M. (2007). The Other White Gold: Salt, Slaves, the Turks and Caicos Islands, and British Colonialism. The Historian. Volume 69, Issue2, Summer 2007. Pages 215-230. (See: https://onlinelibrary.wiley.com/doi/full/10.1111/j.1540-6563.2007.00178.x [Retrieved 7 April 2020].
[158] https://onlinelibrary.wiley.com/doi/full/10.1111/j.1540-6563.2007.00178.x [Retrieved 7 April 2020].
[159] https://turksandcaicostourism.com/about-turks-and-caicos/ [Retrieved 7 April 2020].

[160] http://www.smithsonianmag.co/history/white-gold-how-salt-made-and-unamde-the-turks-and-caicos-islands-161576195/ [Retrieved 7 April 2020].
[161] https://www.britannica.com/place/Turks-and-Caicos-Islands [Retrieved: 17 January 2021].
[162] De Booy, T. (1918). The Turks and Caicos Islands, British West Indies. Geographical Review, 6(1), 37-51. doi:10.2307/207448.
[163] Kennedy, Cynthia M. (2007). The Other White Gold: Salt, Slaves, the Turks and Caicos Islands, and British Colonialism. The Historian. Volume 69, Issue2, Summer 2007. Pages 215-230.
(See: https://onlinelibrary.wiley.com/doi/full/10.1111/j.1540-6563.2007.00178.x [Retrieved: 7 April 2020]..
[164] https://suntci.com/discrediting-the-myth-tci-dependent-or-not-p5671-129.htm [Retrieved: 30 December 2020].
[165] https://suntci.com/discrediting-the-myth-tci-dependent-or-not-p5671-129.htm [Retrieved: 30 December 2020]
[166] https://suntci.com/discrediting-the-myth-tci-dependent-or-not-p5671-129.htm [Retrieved: 30 December 2020].
[167] https://suntci.com/discrediting-the-myth-tci-dependent-or-not-p5671-129.htm [Retrieved: 30 December 2020].
[168] https://bahamasgeotourism.com/entries/morton-salt-company/dc3fe522-9398-4bdf-81da-3e3bd0e697a1 [Retrieved: 17 January 2021].
[169] https://suntci.com/discrediting-the-myth-tci-dependent-or-not-p5671-129.htm [Retrieved: 30 December 2020].
[170] Mills, Carlton (Ed.). (2008). *A History of the Turks and Caicos Islands*. Oxford: MacMillan Publishers. p. 283.
[171] https://www.visittci.com/nature-and-history/history/canada-proposed-union [Retrieved: 17 January 2021].
[172] https://www.visittci.com/nature-and-history/history/population [Retrieved: 17 January 2021].
[173] https://www.visittci.com/nature-and-history/history/population [Retrieved: 17 January 2021].
[174] https://www.britannica.com/place/Turks-and-Caicos-Islands [Retrieved: 20 March 2020].
[175] https://www.visittci.com/nature-and-history/history/population [Retrieved: 17 January 2021].
[176] https://www.visittci.com/nature-and-history/history/population [Retrieved: 17 January 2021].

[177] https://www.visittci.com/nature-and-history/history/population [Retrieved: 17 January 2021].
[178] https://onlinelibrary.wiley.com/doi/full/10.1111/j.1540-6563.2007.00178.x [Retrieved 7 April 2020].
[179] https://www.britannica.com/place/Turks-and-Caicos-Islands [Retrieved 20 April 2020].
[180] https://www.britannica.com/place/Turks-and-Caicos-Islands [Retrieved 20 April 2020].
[181] https://www.visittci.com/providenciales/about [Retrieved 7 April 2020].
[182] https://www.visittci.com/grand-turk-cruise-center [Retrieved 7 April 2020].
[183]http://www.caribbeanelections.com/knowledge/biography/bios/mccartney_james.asp. [Retrieved: 17 January 2021].
[184] Tinker, Keith. (2011). The Migration of Peoples from the Caribbean to the Bahamas. Florida: University Press of Florida, pp.7-8. Note: Many people, who live in The Bahamas are descendants of the Turks and Caicos Islanders
[185]http://www.caribbeanelections.com/knowledge/biography/bios/mccartney_james.asp. [Retrieved: 17 January 2021].
[186] https://www.visittci.com/nature-and-history/history/population [Retrieved: 17 January 2021].
[187]https://en.wikisource.org/wiki/1911_Encyclopædia_Britannica/Antilles [Retrieved 20 March 2020].
[188] Phelan, John Leddy. (1970). *The Millennial Kingdom of the Franciscans in the New World.* California: University of California Press. p. 70.
[189] Albury, Paul. (1975). *The Story of the Bahamas*. MacMillan Caribbean. pp.17-18. See also: Craton, Michael. (1986). *A History of the Bahamas*. San Salvador Press. p. 13.
[190] Risi, Vincenzo De. (2015). *Mathematizing Space: The Objects of Geometry from Antiquity to the Early Modern Age*. New York: Spronger Cham Heidelberg. p. 143.
[191] Markham, Clements R. (Ed.). (2010). *Letters of Amerigo Vespucci, and Other Documents Illustrative of His Career*. New York: Cambridge University Press, pp. xvi-xviii.
[192] https://www.britannica.com/topic/Dutch-West-India-Company [Retrieved: 17 January 2021].

[193] Rosanne Adderly, (2000). *Encyclopedia of Contemporary Latin American and Caribbean Cultures, Volume 1: A-D*. London and New York: Routledge. p.1584.
[194] https://www.history.com/news/us-virgin-islands-denmark-purchase [Retrieved: 17 January 2021].
[195] https://www.globalsecurity.org/military/world/caribbean/sb-history.htm [Retrieved: 17 January 2021].
[196] Smith, Dorsía, Tagirova, Tatiana & Engman, Suzanna. (Eds.). (2010). *Critical Perspectives on Caribbean Literature and Culture*. Newcastle upon Tyne: Cambridge Scholars Publishing. p. 1.
[197] https://www.britannica.com/place/Lesser-Antilles [Retrieved 10 June 2020].
[198] Tinker, Keith. (2012). The African Diaspora to the Bahamas. Victoria, BC, Canada: L. Friesen Press, p. 58.
[199] https://en.wikipedia.org/wiki/Bermuda [Retrieved 15 June 2020].
[200] https://www.britannica.com/place/Bermuda [Retrieved: 10 June 2020].
[201] https://www.bl.uk/anglo-saxons/articles/how-was-the-kingdom-of-england-formed# [Retrieved 15 June 2020].
[202] https://www.britannica.com/place/Wales [Retrieved 15 June 2020].
[203]https://en.wikipedia.org/wiki/Union_of_the_Crowns#:~:text=The%20Union%20of%20the%20Crowns,single%20monarch%20on%2024%20March [Retrieved 15 June 2020].
[204]https://www.thoughtco.com/age-of-exploration-1435006 [Retrieved 15 June 2020].
[205] https://en.wikipedia.org/wiki/Thirteen_Colonies [Retrieved 10 April 2020].
[206] https://www.historic-uk.com/HistoryUK/HistoryofBritain/The-UK-Great-Britain-Whats-the-Difference/ [Retrieved 15 June 2020].
[207] https://www.etymonline.com/search?q=tourist [Retrieved 21 Jan. 2014].
[208] https://www.etymonline.com/search?q=tourist [Retrieved 21 Jan. 2014].
[209] Funk, Wilfred. (1998). Word Origins: An Exploration and History of Words and Language. New York: Wings Books, p.331.
[210] https://www.britannica.com/topic/Carib [Retrieved 20 June 2020].
[211] https://www.britannica.com/place/Pangea [Retrieved 20 June 2020].

[212] https://www.britannica.com/place/Bering-Strait [Retrieved 20 June 2020].
[213] https://www.britannica.com/place/Beringia [Retrieved 20 June 2020].
[214] https://www.britannica.com/biography/Vitus-Bering [Retrieved 20 June 2020].
[215]https://www.merriam-webster.com/dictionary/eastern%20hemisphere [Retrieved: 17 January, 2021].
[216] Wells, Spencer. (2002). The Journey of Man—A Genetic Odyssey. New Jersey: Princeton University Press, pp. 138,142.
[217] Bradley, Bruce and Stanford, Dennis. (2004). The North Atlantic ice-edge corridor: A possible Paleolithic route to the New World. *World Archaeology*, 36(4), 459-478.
[218] "The Human Journey: Migration Routes." Genographic Project. N.p., n.d. [Retrieved 30 Aug. 2013].
[219] Wells, Spencer. (2002). *The Journey of Man—A Genetic Odyssey.* Princeton University Press: New Jersey, p. 138.
[220] "Native American Populations Descend from Three Key Migrations, Scientists Say." ScienceDaily, 11 July 2012. [Retrieved 12 Oct. 2013].
[221] Wells, Spencer. (2002). *The Journey of Man—A Genetic Odyssey.* Princeton University Press: New Jersey, p. 141.
[222] https://www.smithsonianmag.com/science-nature/how-humans-came-to-americas-180973739/ [Retrieved 16 January 2021 2013].
[223] Craton, Michael and Saunders, Gail. (1999). *Islanders in the Stream: A History of the Bahamian People From Aboriginal Times to the End of Slavery.* Georgia: University of Georgia Press, pp.8-9.
[224] Higman, B. W. (2011). *A Concise History of the Caribbean.* New York: Cambridge University Press, pp. 10-12.
[225] Higman, B. W. (2011). *A Concise History of the Caribbean.* New York: Cambridge University Press, pp. 10-12.
[226] Moure, Ramon Dacal and Rivero, Manuel de La Calle. (1996). *Art and Archaeology of Pre-Columbian Cuba.* Pennsylvania: University of Pittsburgh Press, pp. 22-24.
[227] http://www.sci-news.com/archaeology/human-arrival-bahamas-09416.html [Retrieved: 7 March 2021].
[228] https://news.ncsu.edu/2020/01/columbus-caribbean-claims/ [Retrieved 10 April 2020].

[229] https://ammcbahamas.com/ancient-lucayan-skeletons-discovered-in-clarence-town/ [Retrieved 17 January 2021].
[230] https://www.nature.com/articles/s41598-019-56929-3 [Retrieved: 18 April 2020].
[231] https://www.ncbi.nlm.nih.gov/pmc/articles/PMC4275895/ [Retrieved 17 January 2021].
[232] William F. Keegan, Hofman, Corinne L. and Ramos, Reniel Rodriguez. (2013). *The Oxford Handbook of Caribbean Archaeology*. New York: Oxford University Press. p. 1219.
[233] William F. Keegan, Hofman, Corinne L. and Ramos, Reniel Rodriguez. (2013). *The Oxford Handbook of Caribbean Archaeology*. New York: Oxford University Press. p. 126.
[234] Reid, Basil A. (2009). *Myths and Realities of Caribbean History.* Alabama: University of Alabama Press. p. 48.
[235] William F. Keegan, Hofman, Corinne L. and Ramos, Reniel Rodriguez. (2013). *The Oxford Handbook of Caribbean Archaeology*. New York: Oxford University Press. p. 129.
[236] William F. Keegan, Hofman, Corinne L. and Ramos, Reniel Rodriguez. (2013). *The Oxford Handbook of Caribbean Archaeology*. New York: Oxford University Press. p.1 30.
[237] http://islandluminous.fiu.edu/part01-slide02.html [Retrieved 20 April 2020].
[238] http://islandluminous.fiu.edu/part01-slide02.html [Retrieved 20 April 2020].
239 Keegan, William F. and Carlson, Lisabeth A. (2008).*Talking Taino: Caribbean Natural History from a Native Perspective*. Alabama: University of Alabama Press. p. 2.
[240] Allsopp, Richard and Allsopp, Jeanette (Eds.). (2003). Dictionary of Caribbean English Usage. Jamaica: University of the West Indies Press. p.38.
[241] Honychurch, Lennox. (1995). The Caribbean People. Book1. UK: Nelson Caribbean. p. 69.
[242] Meggers, B.J. and Evans, C. (1978). *Lowland South America and the Antilles*. Ancient Native Americans. W. H. Freeman: San Francisco. pp. 543–591.
[243] Reid, Basil A. (2009). *Myths and Realities of Caribbean History.* Alabama: University of Alabama Press. p. 51.
[244]https://www.nature.com/articles/s41598-019-56929-3 [Retrieved 20 May 2020].

[245]https://www.nature.com/articles/s41598-019-56929-3 [Retrieved 15 May 2020].
[246] Keegan, William F., and Hofman, Corinne L. (2017). *The Caribbean before Columbus*. New York: Oxford University Press. p. 53.
[247] Keegan, William F. and Carlson, Lisabeth A. (2008). *Talking Taino: Caribbean Natural History from A Native Perspective*. Alabama: University of Alabama Press. p. 2.
[248]https://www.blackhistorymonth.org.uk/article/section/pre-colonial-history/4235/ [Retrieved 15 May 2020].
[249] Keegan, William F. and Carlson, Lisabeth A. (2008). *Talking Taino: Caribbean Natural History from A Native Perspective*. Alabama: University of Alabama Press. p. 2.
[250] Keegan, William F. & Hofman, Corinne L. (2017). *The Caribbean before Columbus*. New York: Oxford University Press. pp.12-13.
[251] Reid, Basil A. (2009). *Myths and Realities of Caribbean History*. Alabama: University of Alabama Press. p. 49.
[252] https://www.floridamuseum.ufl.edu/histarch/research/haiti/en-bas-saline/taino-culture/ [Retrieved 15 May 2020].
[253] https://www.floridamuseum.ufl.edu/histarch/research/haiti/en-bas-saline/taino-culture/ [Retrieved 10 January 2020].
[254] https://www.floridamuseum.ufl.edu/histarch/research/haiti/en-bas-saline/taino-culture/ [Retrieved 15 May 2020].
[255] https://www.nature.com/articles/s41598-019-56929-3 [Retrieved 15 May 2020].
[256] Keegan, William F. and Carlson, Lisabeth A. (2008). *Talking Taino: Caribbean Natural History from A Native Perspective*. Alabama: University of Alabama Press. pp. 11-12.
[257] Hulme, P. (1993). Making Sense Of The Native Caribbean. NWIG: New West Indian Guide / Nieuwe West-Indische Gids, 67(3/4), 189-220. Retrieved July 3, 2021, from http://www.jstor.org/stable/41849536.
[258] Reid, Basil A. (2009). *Myths and Realities of Caribbean History*. Alabama: University of Alabama Press. pp. 53-54.
[259] Reid, Basil A. (2009). *Myths and Realities of Caribbean History*. Alabama: University of Alabama Press. pp. 53-54.
[260] Keegan, William F., Hofman, Corinne L. and Rodriguez, Reniel Ramos (Eds.). (2013). The Oxford Handbook of Caribbean Archaeology. Oxford: Oxford University Press, p. 11.

[261] Keegan, William F., Hofman, Corinne L. and Rodriguez, Reniel Ramos (Eds.). (2013). T*he Oxford Handbook of Caribbean Archaeology*. Oxford: Oxford University Press, p. 11.
[262] Reid, Basil A. (2009). *Myths and Realities of Caribbean History*. Alabama: University of Alabama Press. p.55.
[263] Keegan, William F. and Carlson, Lisabeth A. (2008). Talking Taino: Caribbean Natural History from a Native Perspective. Alabama: University of Alabama Press, p. 1.
[264] Reid, Basil A. (2009). *Myths and Realities of Caribbean History*. Alabama: University of Alabama Press. p. 57.
[265] Granberry, Julian and Vescelius, Gary (1992). *Languages of the Pre-Columbian Antilles*. University of Alabama Press. pp. 20-21.
[266] "Hispaniola." Genocide Studies Program. Yale University, 2010. [Retrieved 24 Sep. 2013].
[267] https://en.wikipedia.org/wiki/Cacique#Taino_dynasty [Retrieved 10 June 2020].
[268] https://www.floridamuseum.ufl.edu/histarch/research/haiti/en-bas-saline/taino-society/ [Retrieved 10 June 2020].
269 Fewkes, Jesse Walter. (2009). The Aborigines of Puerto Rico and Neighboring Islands. Alabama: The University of Alabama Press, p. 70.
270 Saunders, Nicholas. (Ed.) (2005). The Peoples of the Caribbean: An Encyclopedia of Archaeology and Traditional Culture. California: ABC-CLIO, Inc., p. 149.
[271] Cobley, Alan. (Ed.). (1994). *Crossroads of Empire: The European-Caribbean Connection, 1492-1992*. Barbados: University of the West Indies (Cave Hill, Barbados). pp. 24–36.
[272] Saunders, Nicholas. (Ed.) (2005). The Peoples of the Caribbean: An Encyclopedia of Archaeology and Traditional Culture. California: ABC-CLIO, Inc., p. 161.
[273] Courtz, Henk. (2008). A Carib Grammar and Dictionary. Toronto: Magoria Books, p. 3.
[274] https://ammcbahamas.com/ancient-lucayan-skeletons-discovered-in-clarence-town/ [Retrieved 17 January 2021].
[275] https://www.ncbi.nlm.nih.gov/pmc/articles/PMC4275895/ [Retrieved 17 January 2021]. See also: Keegan, William F., and Hofman, Corinne L. (2017). *The Caribbean before Columbus*. New York: p. 12.
[276] Cobley, Alan. (Ed.). (1994). *Crossroads of Empire: The European-Caribbean Connection, 1492-1992*. Barbados: University of the West Indies (Cave Hill, Barbados). pp. 24–36.

[277]https://www.oxfordhandbooks.com/view/10.1093/oxfordhb/9780195392302.001.0001/oxfordhb-9780195392302-e-1#oxfordhb-9780195392302-bibItem-14 [Retrieved 15 May 2020].
[278] Flores, Lisa Pierce. (2010). *The History of Puerto Rico*. Santa Barbara: Greenwood Press. p.18.
[279] Keegan, William F. and Carlson, Lisabeth A. (2008). *Talking Taino: Caribbean Natural History from A Native Perspective*. Alabama: University of Alabama Press. p. 12.
[280] Boucher, Philip P. (1992). Cannibal Encounters: Europeans and Island Caribs,1492–1763. Maryland: John Hopkins University, p. 1632.
[281] Boucher, Philip P. (1992). Cannibal Encounters: Europeans and Island Caribs, 1492–1763. Maryland: John Hopkins University Press, pp.1632, 1633, 1636, 1638.
[282] Boucher, Philip P. (1992). Cannibal Encounters: Europeans and Island Caribs, 1492–1763. Maryland: John Hopkins University, p. 1632.
[283] Boucher, Philip P. (1992). Cannibal Encounters: Europeans and Island Caribs, 1492–1763. Maryland: John Hopkins University Press, p. 1632.
[284] Saunders, Nicholas. (Ed.). (2005). *The Peoples of the Caribbean: An Encyclopedia of Archaeology and Traditional Culture.* ABC-CLIO, Inc. California, p.45.
[285] Arens, William. (1979). The Man-Eating Myth: Anthropology and Anthropophagy. London: Oxford University Press, p.44.
[286] https://www.floridamuseum.ufl.edu/science/carib-skulls-boost-credibility-of-columbus-cannibal-claims/ [Retrieved 15 May 2020].
[287] Boucher, Philip P. (1992). Cannibal Encounters: Europeans and Island Caribs, 1492–1763. Maryland: John Hopkins University Press. p. 1633.
[288] Boucher, Philip P. (1992). Cannibal Encounters: Europeans and Island Caribs, 1492–1763. Maryland: John Hopkins University Press. pp.1632, 1633, 1636, 1638
[289] Palmié, Stephan and Scarano, Francisco A. (Eds.) (2011).The Caribbean: A History of the Region and Its Peoples. Chicago: University of Chicago Press, p. 27.
[290] The Second Voyage of Christopher Columbus. https://www.thoughtco.com/the-second-voyage-of-christopher-columbus-2136700 [Retrieved 5 May 2020].
[291] Haas, Jonathan. (ed.) (1990). The Anthropology of War. Cambridge: Cabridge University Press. p.150.

[292] Boucher, Philip P. (1992). *Cannibal Encounters: Europeans and Island Caribs,1492–1763*. Maryland: John Hopkins University, p. 1632.
[293] Keegan, William F. and Carlson, Lisabeth A. (2008). Talking Taino: Caribbean Natural History from a Native Perspective. Alabama: University of Alabama Press, pp. 10-12.
[294] Taylor, Chris (2012). The Black Carib Wars: Freedom, Survival, and the Making of the Garifuna. Oxford: Signal Books. p. 6.
[295] https://www.britannica.com/topic/Carib [Retrieved: 20 January 2020].
[296]http://news.gov.dm/news/2258-parliament-approves-indigenous-people-name-change [Retrieved: 20 January 2020].
[297] Anderson, Gerald H. (1999). *Biographical Dictionary of Christian Missions*. Simon & Schuster Macmillan. p. 89.
[298] Taylor, D. (1946). Kinship and Social Structure of the Island Carib. *Southwestern Journal of Anthropology, 2*(2), 180-212. Retrieved May 6, 2020, from www.jstor.org/stable/3628680.
[299] Courtz, Henk. (2008). A Carib Grammar and Dictionary. Toronto: Magoria Books. p. 1. Taylor, Chris. (2012). The Black Carib Wars: Freedom, Survival, and the Making of the Garifuna. Oxford: Signal Books. p. 4.
[300] Taylor, Chris. (2012). The Black Carib Wars: Freedom, Survival, and the Making of the Garifuna. Oxford: Signal Books. p. 4.
[301] Cobley, Alan. (Ed.). (1994). *Crossroads of Empire: The European-Caribbean Connection, 1492-1992*. Barbados: University of the West Indies (Cave Hill, Barbados). pp. 24–36.
[302] Cobley, Alan. (Ed.). (1994). *Crossroads of Empire: The European-Caribbean Connection, 1492-1992*. Barbados: University of the West Indies (Cave Hill, Barbados). pp. 24–36.
[303] Kalinago (Carib) Resistance to European Colonisation of the Caribbean. Caribbean Quarterly. 54 (4): 79. 2008.
[304] Thornton, John and Kelly, John. (1998). *Africa and Africans in the Making of the Atlantic World, 1400-1800*. New York: Cambridge University Press. p. 284.
[305] Thornton, John. (1998). *Africa and Africans in the Making of the Atlantic World, 1400-1800*. New York: Cambridge University Press .pp. 40-41.
[306] Thornton, John (1998). Africa and Africans in the Making of the Atlantic World, 1400-1800. Cambridge: Cambridge University Press. p.288.

[307] Ibid., p. 161.
[308] Taylor, Chris (2012). The Black Carib Wars: Freedom, Survival, and the Making of the Garifuna. Oxford: Signal Books. p. 4.
[309] Saunders, Nicholas. (Ed.) (2005). The Peoples of the Caribbean: An Encyclopedia of Archaeology and Traditional Culture. California: ABC-CLIO, Inc.
[310] http://news.gov.dm/index.php/news/2258-parliament-approves-indigenous-people-name-change [Retrieved 5 May 2020].
[311] Ibid. Courtz, Henk (2008). A Carib Grammar and Dictionary. Toronto: Magoria Books. p. 1.
[312] Cited in: Craton, Michael. (1986) *A History of the Bahamas*. San Salvador Press. p. 21.
[313] https://www.pnas.org/content/115/10/2341 [Retrieved: 22 November 2019].
[314] Keegan, William F., Hofman, Corinne L. and Rodriguez, Reniel Ramos (Eds.). (2013). The Oxford Handbook of Caribbean Archaeology. Oxford: Oxford University Press. p. 11.
[315] Loven, Sven. (2010). Origins of the Tainan Culture, West Indies. Alabama: University of Alabama Press. p. 1.
[316] Loven, Sven. (2010). *Origins of the Tainan Culture, West Indies.* Alabama: University of Alabama Press. p. 71.
[317] Loven, Sven. (2010). Origins of the Tainan Culture, West Indies. Alabama: University of Alabama Press. p.10. See also: Craton, Michael. (1986). A History of the Bahamas. San Salvador Press. p. 17. Albury, Paul. (1975). The Story of the Bahamas. London: MacMillan Caribbean. pp. 5, 13-14.
[318] Antiquity, Volume 63, Issue 239, June 1989, pp. 373 – 379.
[319] Keegan, William F. and Carlson, Lisabeth A. (2008). *Talking Taino: Caribbean Natural History from A Native Perspective.* Alabama: University of Alabama Press. p. 12
[320] Craton, Michael and Saunders, Gail. (1992}. Islanders in the Stream: From aboriginal times to the end of slavery Volume I Georgia: University of Georgia Press. pp.9, 398.
[321] Keegan, William F. and Carlson, Lisabeth A. (2008). *Talking Taino: Caribbean Natural History from A Native Perspective.* Alabama: University of Alabama Press. p. 12.
[322] Keegan, William F., et al. (Eds.). (2013). *The Oxford Handbook of Caribbean Archaeology*. New York: Oxford University Press. p. 264.
[323] https://www.floridamuseum.ufl.edu/histarch/research/haiti/en-bas-saline/taino-culture/ [Retrieved 15 May 2020].

[324] Keegan, William F., et al. (Eds.). (2013). *The Oxford Handbook of Caribbean Archaeology*. New York: Oxford University Press. p. 265.
[325] Keegan, William F., et al. (Eds.). (2013). *The Oxford Handbook of Caribbean Archaeology*. New York: Oxford University Press. p. 265.
[326] Craton, Michael and Saunders, Gail. (1992). *Islanders in the Stream: A History of the Bahamian People from Aboriginal Times to the End of Slavery*. Georgia: University of Georgia Press, p. 18.
[327] Keegan, William F., et al. (Eds.). (2013). *The Oxford Handbook of Caribbean Archaeology*. New York: Oxford University Press. pp. 264-265.
[328] http://www.sci-news.com/archaeology/human-arrival-bahamas-09416.html [Retrieved: 7 March 2021].
[329] Riley, Sandra. (2000). *Homeward Bound: A History of the Bahama Islands to 1850 with a Definitive Study of Abaco In the American Loyalist Plantation Period*. Florida: Riley Hall Publishers. p. 16.
[330] Keegan, William F. and Carlson, Lisabeth A. (2008). *Talking Taino: Caribbean Natural History from a Native Perspective*. Alabama: University of Alabama Press. pp. 7-8.
[331] https://www.timespub.tc/2020/10/lucayan-legacies [Retrieved: 17 January 2021].
[332] https://www.floridamuseum.ufl.edu/caribarch/education/tc-peoples/ [Retrieved: 21 January 2021].
[333] Keegan, William F. and Carlson, Lisabeth A. (2008). *Talking Taino: Caribbean Natural History from a Native Perspective*. Alabama: University of Alabama Press. p. 21.
[334] Craton, Michael. (1986). *A History of the Bahamas*. San Salvador Press. p. 18. By 1985, over 175 sites were identified.
[335] Colón, Fernando. (1992). *The Life of the Admiral Christopher Columbus by His Son Ferdinand*. Translated by Benjamin Keen. New Brunswick: Rutgers University Press. pp 59–60, 64.
[336] Keegan, William F., Hofman, Corinne L. and Rodriguez, Reniel Ramos (Eds.). (2013). *The Oxford Handbook of Caribbean Archaeology*. Oxford: Oxford University Press. p. 20.
[337] Craton, Michael. (1986). *A History of the Bahamas*. San Salvador Press. p. 21.
[338] Albury, Paul. (1975). *The Story of the Bahamas*. MacMillan Caribbean. pp.17-18. See also: Craton, Michael. (1986). *A History of The Bahamas.* San Salvador Press. pp. 14-16.
[339] https://www.timespub.tc/2020/10/lucayan-legacies [Retrieved: 17 January 2021].

[340] Mills, Carlton (Ed.) (2008). *A History of the Turks and Caicos Islands.* Oxford: MacMillan Publishers. pp. 83-84.
[341] Craton, Michael. (1986). *A History of the Bahamas*. San Salvador Press. pp. 19-20, 24.
[342] Keegan, William F. and Carlson, Lisabeth A. (2008). *Talking Taino: Caribbean Natural History from A Native Perspective*. Alabama: University of Alabama Press. pp. 43-44.
[343] https://www.geraceresearchcentre.com/pdfs/14thNatHist/151-172_Berman.pdf [Retrieved 5 May 2020].
[344] Keegan, William F. and Carlson, Lisabeth A. (2008). *Talking Taino: Caribbean Natural History from a Native Perspective*. Alabama University of Alabama Press. p. 5.
[345] Keegan, William F. and Carlson, Lisabeth A. (2008). *Talking Taino: Caribbean Natural History from a Native Perspective*. Alabama University of Alabama Press. p. 5.
[346] Keegan, William F. (1992). *The People Who Discovered Columbus: The Prehistory of the Bahamas*. University Press of Florida. pp. 166-167. Albury, Paul. (1975). *The Story of the Bahamas*. MacMillan Caribbean. pp. 17-18. Craton, Michael. (1986). *A History of the Bahamas*. San Salvador Press. p. 23.
[347] Craton, Michael. (1986). *A History of the Bahamas*. San Salvador Press. p. 24.
[348] Craton, Michael. (1986). *A History of the Bahamas*. San Salvador Press. p. 18. By 1985, over 175 sites were identified.
[349] Albury, Paul. (1975). *The Story of the Bahamas*. MacMillan Caribbean. pp. 17-18. See also: Craton, Michael. (1986). *A History of the Bahamas*. San Salvador Press. p.17.
[350] Craton, Michael. (1986). *A History of the Bahamas*. San Salvador Press. p. 24.
[351] Craton, Michael. (1986). *A History of the Bahamas*. San Salvador Press. p. 11.
[352] Conrad, Geoffrey W., John W. Foster, and Charles D. Beeker. (2001). "Organic artifacts from the Manantial de la Aleta, Dominican Republic: preliminary observations and interpretations", *Journal of Caribbean Archaeology*. *2*(6). See also: Albury, Paul. (1975). *The Story of the Bahamas*. MacMillan Caribbean. pp.17-18. Craton, Michael. (1986). *A History of the Bahamas*. San Salvador Press. p. 17.

[353] https://www.washingtonpost.com/history/2019/10/14/here-are-indigenous-people-christopher-columbus-his-men-could-not-annihilate/ [Retrieved 20 April 2020].
[354] Craton, Michael. (1986). *A History of the Bahamas*. San Salvador Press. pp. 24, 26.
[355] Keegan, William F. and Carlson, Lisabeth A. (2008). *Talking Taino: Caribbean Natural History from A Native Perspective*. Alabama: University of Alabama Press. p. 4.
[356] Saunders, Nicholas. (Ed.) (2005). *The Peoples of the Caribbean: An Encyclopedia of Archaeology and Traditional Culture.* ABC-CLIO, Inc.: California. pp. 169, 405.
[357] Craton, Michael. (1986). *A History of the Bahamas*. San Salvador Press. pp. 20, 24.
[358] Keegan, William F. (1992). *The People Who Discovered Columbus: The Prehistory of the Bahamas*. University Press of Florida. pp.124-127. Craton, Michael. (1986). *A History of the Bahamas*. San Salvador Press. pp. 20, 25.
[359] https://www.timespub.tc/2020/10/lucayan-legacies [Retrieved: 17 January 2021].
[360] Albury, Paul. (1975). *The Story of the Bahamas*. MacMillan Caribbean. pp. 17-18. Craton, Michael. (1986). *A History of the Bahamas*. San Salvador Press. pp. 20, 24-25. Granberry, Julian & Gary S. Vescelius. (2004). *Languages of the Pre-Columbian Antilles*. The University of Alabama Press. p. 43. Keegan, William F. (1992). *The People Who Discovered Columbus: The Prehistory of the Bahamas.* University Press of Florida. pp. 52-53, 77.
[361] Craton, Michael and Saunders, Gail. (1999). *Islanders in the Stream: A History of the Bahamian People-From Aboriginal Times to the End of Slavery*. Georgia: University of Georgia Press. p. 10.
[362] Craton, Michael and Saunders, Gail. (1999). *Islanders in the Stream: A History of the Bahamian People-From Aboriginal Times to the End of Slavery*. Georgia: University of Georgia Press. p. 10.
[363] Craton, Michael and Saunders, Gail. (1992). *Islanders in the Stream: A History of the Bahamian People: Volume One: From Aboriginal Times to the End of Slavery*. Georgia: The University of Georgia Press. p. 27.
[364] Craton, Michael. (1986). *A History of the Bahamas*. San Salvador Press. p. 21.
[365] Craton, Michael. (1986). *A History of the Bahamas*. San Salvador Press. p. 21.

[366] Craton, Michael. (1986). *A History of the Bahamas*. San Salvador Press. p. 23.
[367] Keegan, William F. and Carlson, Lisabeth A. (2008). *Talking Taino: Caribbean Natural History from A Native Perspective*. Alabama: University of Alabama Press. p. 13.
[368] Keegan, William F. and Carlson, Lisabeth A. (2008). *Talking Taino: Caribbean Natural History from A Native Perspective*. Alabama: University of Alabama Press. p.13.
[369] Riley, Sandra. (2000). *Homeward Bound: A History of the Bahama Islands to 1850 with a Definitive Study of Abaco In the American Loyalist Plantation Period*. Florida: Riley Hall Publishers. p. 13.
[370] https://www.britannica.com/place/North-America/The-lowlands [Retrieved: 2 January 2021].
[371] https://www.history.com/topics/exploration/leif-erickson-vs-christopher-columbus-video [Retrieved 7 January 2021]
[372] https://www.history.com/topics/exploration/leif-eriksson [Retrieved: 2 January 2021].
[373] https://www.history.com/topics/exploration/leif-eriksson [Retrieved: 2 January 2021].
[374] https://en.wikipedia.org/wiki/Zheng_He [Retrieved: 2 January 2021].
[375] Fritze, Ronald H. (2009). Invented Knowledge: False History, Fake Science and Pseudo-Religions. London, England: Reaktion Books. pp. 96–103.
[376] Kolesnikov-Jessop, Sonia. "Did Chinese Beat out Columbus?" Arts. New York Times, 25 June 2005. [Retrieved 24 Sept. 2013].
[377] Menzies, Gavin. (2004). *1421: The Year China Discovered America.* New York: HarperCollins Publishers Inc. p. 303.
[378] Belling, Larry (Actor) and Wallace, David (Director) (2004). 1421: The Year China Discovered America? PBS Documentary. See: Andrea, Alfred and Overfield, James. (2011). The Human Record: Sources of Global History, Volume I: To 1500. Boston: Cengage Learning. p. 408.
[379] Joseph, Frank and Sitchin, Zecharia. (Eds.). (2006). Discovering the Mysteries of Ancient America: Lost History and Legends Unearthed and Explored. Career Press, Inc. pp. 210-217.
[380] Joseph, Frank and Sitchin, Zecharia. (Eds.). (2006). Discovering the Mysteries of Ancient America: Lost History and Legends Unearthed and Explored. Career Press, Inc. pp. 210-217.
[381] Pencak, William. (2011). Historical Dictionary of Colonial America. Maryland: Scarecrow Press, Inc. p. xiv.

[382] Figueredo, D. H. and Argote-Freyre, Frank. (2008). A Brief History of the Caribbean. New York: Facts On File, Incorporated. p.22.
[383] Masur, Louis P. (ed.) (1999). *The Challenge of American History*. Maryland: John Hopkins University Press. p. 31.
[384] https://en.wikipedia.org/wiki/Greenland [Retrieved: 17 January 2021].
[385] https://www.britannica.com/place/East-Indies [Retrieved: 17 January 2021].
[386] Aram, Bethany. (2006). Monarchs of Spain in Iberia and the Americas. Vol. 2.California: ABC Clio. p. 725.
[387] Kohen, Elizabeth, Elias and Marie Louise.(2013). Spain. New York: Marshall Cavendish Benchmark. pp. 19-21.
[388] https://www.nationalgeographic.com/history/article/who-were-moors [Retrieved: 17 January 2021].
[389] Kohen, Elizabeth, Elias and Marie Louise. (2013). Spain. New York: Marshall Cavendish Benchmark. pp. 19-21.
[390] Hazen, Walter. (2005). *Exploration and Discovery*. California: Good Year Books. p. 26.
[391] Smith, Jean Reeder and Smith, Lacey Baldwin (1980). Essentials of World History. London: Barron Educational Series. pp. 76-77.
[392] Kamen, Henry. (2005). *Spain 1469–1714* (3rd ed.). New York: Pearson/Longman. p. 29.
[393] https://www.history.com/topics/exploration/exploration-of-north-america [Retrieved 5 April 2020].
[394] Zinn, Howard. (2015). *A People's History of the United States*. New York: HarperCollins Publishers. pp. 2-4.
[395] Webster, Noah Jr. (1802). *An American Selection of Lessons in Reading and Speaking*. New York: G. & R. Waite. p. 118.
[396] Webster, Noah Jr. (1802). *An American Selection of Lessons in Reading and Speaking*. New York: G. & R. Waite. p. 118.
[397] The Spice Islands and the Age of Exploration. https://brewminate.com/the-spice-islands-and-the-age-of-exploration/ [Retrieved 5 April 2020].
[398] The Spice Islands and the Age of Exploration. https://brewminate.com/the-spice-islands-and-the-age-of-exploration/ [Retrieved 5 April 2020].
[399] https://www.britannica.com/biography/Kublai-Khan [Retrieved 5 April 2020].
[400] The Spice Islands and the Age of Exploration. https://brewminate.com/the-spice-islands-and-the-age-of-exploration/ [Retrieved 5 April 2020].

[401] The Spice Islands and the Age of Exploration. https://brewminate.com/the-spice-islands-and-the-age-of-exploration/ [Retrieved 5 April 2020].
[402] Bartolomeu Dias. https://www.history.com/topics/exploration/bartolomeu-dias [Retrieved 15 May 2020].
[403] https://www.pbs.org/wgbh/aia/part1/1narr1.html [Retrieved 17 January 2021].
[404] https://www.biography.com/explorer/henry-the-navigator [Retrieved 5 May 2020].
[405] https://www.pbs.org/wgbh/aia/part1/1narr1.html [Retrieved 17 January 2021].
[406]https://www.bbc.co.uk/worldservice/africa/features/storyofafrica/9chapter3.shtml [Retrieved 17 January 2021].
[407]https://www.dw.com/en/east-africas-forgotten-slave-trade/a-50126759 [Retrieved 15 May 2020].
[408] https://bbc.co.uk/history/british/abolition/africa_article_01.shtml [Retrieved: 17 January 2021].
[409] Davis, Robert C. (2003). Christian Slaves, Muslim Masters: White Slavery in the Mediterranean, the Barbary Coast and Italy, 1500-1800. New York: Palgrave Macmillan. pp.3-27.
[410] https://www.biography.com/explorer/henry-the-navigator [Retrieved 5 May 2020].
[411] https://www.history.com/topics/exploration/bartolomeu-dias [Retrieved 15 May 2020].
[412] https://www.history.com/topics/exploration/bartolomeu-dias [Retrieved 15 May 2020].
[413] https://www.history.com/topics/exploration/christopher-columbus#section_1 [Retrieved 20 April 2020].
[414] *The Geographical Journal, 136*(2). (Jun., 1970). London: The Royal Geographical Society. pp. 177-189.
[415] https://www.biography.com/explorer/christopher-columbus [Retrieved 20 April 2020].
[416] https://www.biography.com/explorer/christopher-columbus [Retrieved 20 April 2020].
[417] Webster, Noah Jr. (1802). *An American Selection of Lessons in Reading and Speaking*. New York: G. & R. Waite. p. 119.
[418] The Spice Islands and the Age of Exploration. https://brewminate.com/the-spice-islands-and-the-age-of-exploration/ [Retrieved 5 April 2020].

[419] https://www.sjsu.edu/faculty/watkins/theconquest.htm [Retrieved 5 April 2020].
[420] Hazen, Walter. (2005). *Exploration and Discovery*. California: Good Year Books. pp. 27-28.
[421] https://www.history.com/this-day-in-history/columbus-lands-in-south-america [Retrieved 17 January 2021].
[422]http://www.bbc.co.uk/history/british/tudors/columbus_legacy_01.shtml [Retrieved 5 April 2020].
[423] Webster, Noah Jr. (1802). *An American Selection of Lessons in Reading and Speaking*. New York: G. & R. Waite. p.121.
[424] Christopher Columbus Biography--(c. 1451–1506). See also: https://www.biography.com/explorer/christopher-columbus. [Retrieved 15 May 2020].
[425] https://www.britannica.com/biography/Christopher-Columbus [Retrieved 15 May 2020].
[426] Pletcher, Kenneth. (Ed.) (2010). *The Britannica Guide to Explorers and Explorations That Changed the Modern World*. New York: Educational Publishing Britannica Educational Publishing. pp. 60-61.
[427] Bourne, E. G. (Ed.). (1906). *The Northmen, Columbus and Cabot, 985-1503: The voyages of the Northmen, The voyages of Columbus and of John Cabot*. New York: Charles Scribner's Sons. p. 90.
[428] Zinn, Howard. (2015). *A People's History of the United States*. New York: HarperCollins Publishers. pp. 2-4.
[429] Bourne, E. G. (Ed.). (1906). *The Northmen, Columbus and Cabot, 985-1503: The voyages of the Northmen, The voyages of Columbus and of John Cabot*. New York: Charles Scribner's Sons. p. 79.
[430] Sauer, Carl Ortwin. (1966). *The Early Spanish Main*. New York: Cambridge University Press. pp. 24-25.
[431] Haase, Wolfgang and Reinhold, Meyer. (1993). *The Classical Tradition and the Americas: European images of the Americas and The Classical Tradition*. Berlin: Walter De Gruyter. pp. 201-202.
[432] https://www.britannica.com/biography/Roderick [Retrieved 5 June 2020].
[433] Phelan, John Leddy (1970). *The Millennial Kingdom of the Franciscans in the New World*. Los Angeles: University of California Press. p. 70.
[434] Haase, Wolfgang and Reinhold, Meyer. (1993). *The Classical Tradition and the Americas: European images of the Americas and The Classical Tradition*. Berlin: Walter De Gruyter. p. 202.

[435] Phelan, John Leddy. (1970). *The Millennial Kingdom of the Franciscans in the New World.* California: University of California Press. p. 70.
[436] https://www.biography.com/explorer/christopher-columbus [Retrieved March 2020].
[437] http://www.franciscan-archive.org/columbus/opera/excerpts.html [Retrieved March 2020].
[438] Markham, Clements R. (Ed.). (1893). *The Journal of Christopher Columbus (during His First Voyage, 1492–93)*. London: The Hakluyt Society. p. 35.
[439] Riley, Sandra. (2000). *Homeward Bound: A History of the Bahama Islands to 1850 with a Definitive Study of Abaco In the American Loyalist Plantation Period*. Florida: Riley Hall Publishers. p. 5.
[440] Craton, Michael and Saunders, Gail. (1999). *Islanders in the Stream: A History of the Bahamian People-from Aboriginal Times to the End of Slavery*. Georgia: University of Georgia Press. p. 21.
[441] https://www.gilderlehrman.org/sites/default/files/inline-pdfs/01427_fps.pdf [Retrieved 5 June 2020].
[442] Curet L. Antonio and Hauser, Nark W. (2011). *Islands at the Crossroads: Migration, Seafaring, and Interaction in the Caribbean.* Alabama: Alabama Press. p. 119.
[443] Sauer, Carl Ortwin. (1966). *The Early Spanish Main*. New York: Cambridge University Press. pp. 24-25.
[444] Fernández-Armesto, Felipe.(2008). Columbus on Himself. Indiana: Hackett Publishing. p.112.
[445] Fernández-Armesto, Felipe.(2008). Columbus on Himself. Indiana: Hackett Publishing, Inc., pp. 69, 70.
[446] Craton, Michael and Saunders, Gail. (1992). *Islanders in the Stream: A History of the Bahamian People-From Aboriginal Times to the End of Slavery.* Georgia: University of Georgia Press. p. 47.
[447] Excerpt from Columbus' letter to the Spanish Monarchs upon returning from his first voyage. See: https://www.washingtonpost.com/news/retropolis/wp/2018/06/15/the-journey-of-a-hijacked-christopher-columbus-letter-recounting-his-voyage-to-america/ [Retrieved 18 April 2020].
[448] Craton, Michael and Saunders, Gail (1999). *Islanders in the Stream: A History of the Bahamian People-From Aboriginal Times to the End of Slavery*. Georgia: University of Georgia Press. p. 405.
[449] http://www.thebahamasweekly.com/publish/bahamas-historical-society/Lucayan_Topynyms9510.shtml [Retrieved 5 June 2020].

[450] Beding, Silvio A. (Ed.). (1992). *The Christopher Columbus Encyclopedia.* New York: Simon & Schuster Inc. p.402. See also: http://www.thebahamasweekly.com/publish/bahamas-historical-society/Lucayan_Topynyms9510.shtml [Retrieved 5 June 2020].
[451] Craton, Michael and Saunders, Gail. (1999). *Islanders in the Stream: From Aboriginal Times to the End of Slavery. Volume 1.* Athens: University of Georgia Press. pp. 50-51.
[452] Colón, Fernando. (1992). The Life of the Admiral Christopher Columbus by His Son Ferdinand. (Translated by Benjamin Keen). New Brunswick: Rutgers University Press. pp 59–60, 64.
[453] Keegan, William F. and Carlson, Lisabeth A. (2008). *Talking Taino: Caribbean Natural History from A Native Perspective.* Alabama: University of Alabama Press. p. 43.
[454] Keegan, William F. and Carlson, Lisabeth A. (2008). *Talking Taino: Caribbean Natural History from A Native Perspective.* Alabama: University of Alabama Press. p. 10.
[455] Keegan, William F. and Carlson, Lisabeth A. (2008). *Talking Taino: Caribbean Natural History from A Native Perspective.* Alabama: University of Alabama Press. p. 40.
[456] https://www.timespub.tc/2020/10/lucayan-legacies [Retrieved: 17 January 2021].
[457] Keegan, William F. and Carlson, Lisabeth A. (2008). *Talking Taino: Caribbean Natural History from a Native Perspective.* Alabama: University of Alabama Press. pp. 7-8.
[458] The British Crown had granted the Bahama Islands to Attorney General Sir Robert Heath. Craton, Michael. (1986). *A History of the Bahamas.* Ontario, Canada: San Salvador Press, p. 50.
[459] Craton, Michael and Saunders, Gail. (1999). *Islanders in the Stream: A History of the Bahamian People: Volume One: From Aboriginal Times to the End of Slavery.* Athens, Georgia: The University of Georgia Press, pp.74-78.
[460] Craton, Michael. (1986). *A History of the Bahamas.* San Salvador Press. p. 21.
[461] https://www.latimes.com/archives/la-xpm-2010-feb-01-la-oe-millerzinn1-2010feb01-story.html [Retrieved: 20 January 2020].
[462] https://www.nature.com/articles/s41598-019-56929-3 [Retrieved 15 May 2020].
[463] Letter of Christopher Columbus on His First Voyage to America, 1492.

https://nationalhumanitiescenter.org/pds/amerbegin/contact/text1/columbusletter.pdf [Retrieved 5 May 2020].
[464] https://www.washingtonpost.com/history/2019/10/14/here-are-indigenous-people-christopher-columbus-his-men-could-not-annihilate/ [Retrieved 20 April 2020].
[465] Craton, Michael and Saunders, Gail. (1999). *Islanders in the Stream: A History of the Bahamian People-From Aboriginal Times to the End of Slavery*. Georgia: University of Georgia Press. p. 7.
[466]Keegan, William F. and Carlson, Lisabeth A. (2008) Talking Taino: Caribbean Natural History from a Native Perspective. Alabama: University of Alabama Press, p.7.
[467] Figueredo, D. H. and Argote-Freyre, Frank. (2008). A Brief History of the Caribbean. New York: Facts On File, Incorporated. p.22.
[468] https://news.ncsu.edu/2020/01/columbus-caribbean-claims/ [Retrieved 10 April 2020].
[469] Loven, Sven. (2010). Origins of the Tainan Culture, West Indies. Alabama: University of Alabama Press. pp. 57-58.
[470] Craton, Michael and Saunders, Gail. (1992). *Islanders in the Stream: A History of the Bahamian People-From Aboriginal Times to the End of Slavery*. Georgia: University of Georgia Press. p. 39.
[471] Keegan, William and Hofman, Corinne. (2017).*The Caribbean before Columbus*. New York: Oxford University Press. p. 242.
[472] https://www.floridamuseum.ufl.edu/science/carib-skulls-boost-credibility-of-columbus-cannibal-claims/ [Retrieved 20 May 2020].
[473] https://www.nature.com/articles/s41598-019-56929-3 [Retrieved 15 May 2020].
[474] https://www.nature.com/articles/s41598-019-56929-3 [Retrieved 18 April 2020].
[475] https://news.ncsu.edu/2020/01/columbus-caribbean-claims/ [Retrieved 10 April 2020].
[476] https://www.floridamuseum.ufl.edu/science/carib-skulls-boost-credibility-of-columbus-cannibal-claims/ [Retrieved 20 May 2020].
[477] https://www.floridamuseum.ufl.edu/science/carib-skulls-boost-credibility-of-columbus-cannibal-claims/ [Retrieved 20 May 2020].
[478] https://www.archaeology.org/news/9344-201228-caribbean-migration-dna [Retrieved: 2 January 2021].
[479] https://www.archaeology.org/news/9344-201228-caribbean-migration-dna [Retrieved: 2 January 2021].

[480] https://www.sapiens.org/biology/indigenous-caribbean/ 5 February 2021].
[481] https://www.archaeology.org/news/9344-201228-caribbean-migration-dna [Retrieved: 2 January 2021].
[482] https://arstechnica.com/science/2020/12/facial-profiling-ancient-dna-tell-two-tales-of-early-caribbean-islanders/ [Retrieved 17 January 2021].
[483] https://www.archaeology.org/news/9344-201228-caribbean-migration-dna [Retrieved: 2 January 2021].
[484] https://www.nationalgeographic.org/thisday/oct12/columbus-makes-landfall-caribbean/ [Retrieved 20 March 2020].
[485] https://www.nationalgeographic.org/thisday/oct12/columbus-makes-landfall-caribbean/ [Retrieved 20 March 2020].
[486] Craton, Michael and Saunders, Gail (1999). *Islanders in the Stream: A History of the Bahamian People-From Aboriginal Times to the End of Slavery*. Georgia: University of Georgia Press. p. 405.
[487] Craton, Michael and Saunders, Gail (1999). *Islanders in the Stream: A History of the Bahamian People-From Aboriginal Times to the End of Slavery*. Georgia: University of Georgia Press. p. 405.
[488] The 17th century British pirate, George Watling, after whom the island was named, is said to have claimed the island for himself. See: Parker, Christopher. (2001). *Bahamas and Turks and Caicos.* (2nd ed.). Victoria, Australia: Lonely Planet Publications. p. 373.
[489] Dupuch, Etienne Jr. (2003). Bahamas Handbook. Nassau: Etienne Dupuch Jr. Publications. pp. 67, 72.
[490] Craton, Michael and Saunders, Gail (1999). *Islanders in the Stream: A History of the Bahamian People-From Aboriginal Times to the End of Slavery.* Georgia: University of Georgia Press. p. 405.
[491] Craton, Michael and Saunders, Gail (1999). *Islanders in the Stream: A History of the Bahamian People-From Aboriginal Times to the End of Slavery*. Georgia: University of Georgia Press. p. 405.
[492] McKinnen, Daniel. (1804). *A Tour Through the British West Indies, in the Years 1802 and 1803, Giving a Particular Account of The Bahama Islands*. London: Taylor Black-Horse Court. pp. 198-201.
[493] https://www.loc.gov/resource/g3291s.ct000341/ [Retrieved 10 July 2020].
[494] McKinnen, Daniel. (1804). A Tour Through the British West Indies, in the Years 1802 and 1803, Giving a Particular Account of The Bahama Islands. London: Taylor Black-Horse Court. pp. 198-201.
[495] Each year, hundreds of cruise ship passengers visit Half Moon Cay at Little San Salvador for relaxation and recreation. Dupuch, Etienne Jr.

(2003). Bahamas Handbook. Nassau: Etienne Dupuch Jr. Publications. p. 68.

[496] Dupuch, Etienne, Jr. (1992). Bahamas Handbook and Businessman's Annual. Nassau: E. Dupuch, Jr., Publications. p. 31.

[497] Gerace, Donald T. Publications: The First San Salvador conference; Columbus and his World. Gerace Research Centre, 1987. [Retrieved 19 Apr. 2014].

[498] Saunders, Nicholas. (Ed.) (2005). *The Peoples of the Caribbean: An Encyclopedia of Archaeology and Traditional Culture.* ABC-CLIO, Inc.: California. pp. 169, 405.

[499] Becher, A. (1856). The Landfall of Columbus on His First Voyage to America. *The Journal of the Royal Geographical Society of London, 26,* 189-203. doi:10.2307/1798355.

[500] Dupuch, Etienne Jr. (2003). *Bahamas Handbook.* Nassau: Etienne Dupuch Jr. Publications. p. 65.

[501] Barratt, Peter. (2011). Bahama Saga: The Epic Story of the Bahama Islands. Indiana: 1stBooks. p. 328.

[502] Craton, Michael and Saunders, Gail (1999). *Islanders in the Stream: A History of the Bahamian People-From Aboriginal Times to the End of Slavery.* Georgia: University of Georgia Press. p. 405.

[503] https://www.nationalgeographic.org/thisday/oct12/columbus-makes-landfall-caribbean/ [Retrieved 20 March 2020].

[504] Keegan, William F. and Carlson, Lisabeth A. (2008). *Talking Taino: Caribbean Natural History from a Native Perspective.* Alabama: University of Alabama Press. p. 130.

[505] Https://www.sun-sentinel.com/news/fl-xpm-1990-10-07-9002180406-story.html [Retrieved: 15 March 2020].

[506] https://en.wikipedia.org/wiki/Juan_de_la_Cosa [Retrieved: 17 January 2021].

[507] https://www.fundacionnaovictoria.org/history-nao-santa-maria/ [Retrieved: 17 January 2021].

[508] https://www.britannica.co/mtopic/Santa-Maria-ship [Retrieved: 17 January 2021].

[509] https://www.history.com/news/christopher-columbus-ships-caravels [Retrieved: 17 January 2021].

[510]https://www.nationalgeographic.com/adventure/article/columbus-nina-pinta-santa-maria-shipwreck-archaeology [Retrieved: 17 January 2021].

[511] Bodmer, Beatriz Pastor. (1992). The Armature of Conquest: Spanish Accounts of the Discovery of America, 1492 – 1589. California: Stanford University Press. p. 24.

[512] http://www.keyshistory.org/cuba.html [Retrieved 20 June 2020].
[513] Crooker, Richard A. (2003). *Cuba: Modern World Nations.* Pennsylvania: Chelsea House Publishers. p.57.
[514] https://www.chicagotribune.com/news/ct-xpm-2004-08-10-0408100264-story.html [Retrieved 20 April 2020].
515 Keegan, William F. and Carlson, Lisabeth A. (2008). Talking Taino: Caribbean Natural History from A Native Perspective. Alabama: University of Alabama Press. p. 12.
516 https://www.britannica.com/place/Cuba/Soils#ref515644 [Retrieved 20 April 2020].
517 Keegan, William F. and Carlson, Lisabeth A. (2008). Talking Taino: Caribbean Natural History from A Native Perspective. Alabama: University of Alabama Press. p. 12.
[518] https://de.reuters.com/article/us-cuba-columbus/no-cocktail-for-columbus-at-cuba-landing-monument-idUSTRE57G5JC20090817 [Retrieved 20 April 2020].
[519] Craton, Michael. (1986). *A History of the Bahamas.* San Salvador Press. pp. 14-20.
[520] Craton, Michael. (1986). *A History of the Bahamas.* San Salvador Press. p. 23.
[521] https://www.etymonline.com/word/haiti [Retrieved 10 April 2020].
[522] Hume, Robert. (1992). England: Gracewing Books. Christopher Columbus and the European Discovery of America. p.46.
[523] Sauer, Carl Ortwin. (1966). *The Early Spanish Main.* New York: Cambridge University Press. pp. 24-25.
[524] http://www.ems.kcl.ac.uk/content/pub/b001.html [Retrieved 20 April 2020].
[525] Lavery, Brian. (2013). *The Conquest of the Ocean.* New York: DK Publishing. p. 70.
[526] Saunders, Nicholas. (Ed.) (2005). The Peoples of the Caribbean: An Encyclopedia of Archaeology and Traditional Culture. ABC-CLIO, Inc. California. p. 157.
[527] https://washingtonpost.com/history/2019/1014/here-are-indigenous-people-christopher-columbus-his-men-could-not-annihilate [Retrieved 5 April 2020].
[528]https://academics.hamilton.edu/governmentdparis/govt375/spring98/multiculturism/history/columbus.html [Retrieved 5 April 2020].
[529] Excerpt from Columbus' letter to the Spanish monarchs upon returning from his first voyage. See:

https://www.washingtonpost.com/news/retropolis/wp/2018/06/15/the-journey-of-a-hijacked-christopher-columbus-letter-recounting-his-voyage-to-america/ [Retrieved 18 April 2020].
[530] Zinn, Howard. (2015). *A People's History of the United States*. New York: HarperCollins Publishers. pp. 2-4.
[531] https://www.sjsu.edu/faculty/watkins/theconquest.htm [Retrieved 5 April 2020].
[532] Rodriguez, Junius P. (1997). *The Historical Encyclopedia of World Slavery, Volume 1; Volume 7*. California: ABC-CLIO. p. 24.
[533] https://www.britannica.com/event/Treaty-of-Tordesillas [Retrieved 5 April 2020].
[534] https://www.britannica.com/event/Treaty-of-Tordesillas [Retrieved 5 April 2020].
[535] https://www.britannica.com/event/Treaty-of-Tordesillas {Retrieved: 17 April 2021].
[536] https://www.britannica.com/event/Treaty-of-Tordesillas [Retrieved 5 April 2020].
[537] Rodriguez, Junius P. (1997). *The Historical Encyclopedia of World Slavery, Volume 1; Volume 7*. California: ABC-CLIO. p. 639.

Bibliography

Apperson, George Latimer; Manser, Martin H.; and Stephen J. Curtis. (2006). *Dictionary of Proverbs.* London: Woodsworth Editions Limited.

Anderson, Gerald H. (1999). *Biographical Dictionary of Christian Missions.* Simon & Schuster Macmillan.

Aram, Bethany. (2006). Monarchs of Spain in Iberia and the Americas. Vol. 2.California: ABC Clio.

Bethell, Talbot. (1937). *The Early Settlers of the Bahamas and Colonists of North America.* Maryland: Heritage Books.

Brinkbäumer Klaus, Höges Clemens, Streck Annette. (199). *The Voyage of the Vizcaína: The Mystery of Christopher Columbus' Last Ship.* Florida: Harcourt Inc.

Cash, Phillip and Maples, Don. (1987). *The Making of The Bahamas: A History for Schools.* England: Longman Caribbean.

Crooker, Richard A. (2003). Cuba: Modern World Nations. Pennsylvania: Chelsea House Publishers.

Davis, Robert C. (2003). Christian Slaves, Muslim Masters: White Slavery in the Mediterranean, the Barbary Coast and Italy, 1500-1800. New York: Palgrave Macmillan.

Diffie, Bailey W. and Winius, George D. (1977). *Foundations of the Portuguese empire.* Minnesota: University of Minnesota Press.

Dunan, Marcel. (1964). *Larousse Encyclopedia of Modern History, From 1500 to the Present Day.* New York: Harper and Row.

Figueredo, D. H. and Argote-Freyre, Frank. (2008). A Brief History of the Caribbean. New York: Facts On File, Incorporated. p.22.

Fritze, Ronald H. (2009). Invented Knowledge: False History, Fake Science and Pseudo-Religions. London, England: Reaktion Books.

Granberry, Julian and Vescelius, Gary S. (2004). *Languages of the Pre-Columbian Antilles.* Alabama: The University of Alabama Press.

Green, Edmund, Ravilious, Corinna, & Spalding, Mark. (2001). *World Atlas of Coral Reefs.* Berkeley, CA: University of California Press and UNEP/WCMC.

Haas, Jonathan. (ed.) (1990). The Anthropology of War. Cambridge: Cabridge University Press.

Honychurch, Lennox. (1995). The Caribbean People. Book1. UK: Nelson Caribbean.

Keegan, William F. & Hofman, Corinne L. (2017). *The Caribbean before Columbus.* New York: Oxford University Press.

Lefroy, Sir John, Henry. (1879). Memorials of the Discovery and Early Settlement of the Bermudas Or Somers Island (1511 – 1687), Volume 2. London: Longmans, Green & Co.

Mizell, Louis R. Jr. (1997). *Masters of Deception: The Worldwide White-collar Crime Crisis and Ways to Protect Yourself.* New York: John Wiley and Sons.

McGlynn, Margaret and Bartlett, Kenneth R. (Eds.). (2014). *The Renaissance and Reformation in Northern Europe.* Canada: University of Toronto Press.

National Maritime Historical Society & Sea History Magazine. (Nov 30, 2011). Sea History 137. Winter 2011-2012. New York.

Olmstead, Kathleen. (2008). *Jacques Cousteau: A Life Under the Sea.* New York: Sterling Publishing Co. Inc.

Parry, Dan. (2006). Blackbeard: The Real Pirate of the Caribbean. North Carolina: National Maritime Museum.

Polomé Edgar C., & Winter, Werner. (1992). *Reconstructing Languages and Cultures*. New York: Walter de Gruyter & Co.

Ribiero, Michele D. (Ed.). (2020). *Examining Social Identities and Diversity Issues in Group Therapy: Knocking at the Boundaries.* New York: Routledge Taylor & Francis Group.

Riley, Sandra. (2000). *Homeward Bound: A History of the Bahama Islands to 1850 with a Definitive Study of Abaco In the American Loyalist Plantation Period*. Florida: Riley Hall Publishers.

Roorda, Eric, Paul. (2016). Historical Dictionary of the Dominican Republic. Maryland: Rowman & Littlefield Publishers. p.111.

Sargent, Epes. (1873). *A School Manual of English Etymology: And Text-book of Derivatives, Prefixes and suffixes*. Florida: J. H. Butler & Company.

Sauer, Carl Ortwin (1966). *The Early Spanish Main*. Los Angeles: University of California Press.

Smith, Jean Reeder and Smith, Baldwin Lacey. (1980). *Essentials of World History*. New York: Barron's Educational Series, Inc.

Smyth, William Henry. (2007). *The Sailor's Word: A Complete Dictionary of Nautical Terms from the Napoleonic and Victorian Navies*. Arizona: Fireship Press.

Stausberg, Michael (2011). *Religion and Tourism: Crossroads, Destinations and Encounters*. New York: Routledge.

Taylor, Chris (2012). *The Black Carib Wars: Freedom, Survival, and the Making of the Garifuna*. Oxford: Signal Book.

The Bahamas Government. (1973). *The Bahamas Independence Order*. Nassau, Bahamas: Bahamas Government Printing.

The Ministry of Transport and Aviation. (2015). *The National Maritime Policy*. Nassau: The Commonwealth of The Bahamas.

Turrell, Todd. T. (2015). *History of The Bahamas through Maps.* Florida: Turrell, Hall and Associates. p.34.

Wilchcombe, Obediah (Producer) & Sealy, Gina (Director). (2014). *Gentle Giant: The André Rodgers Story.* Bahamas.

Woodard, Colin. (2008). *The Republic of Pirates: Being the True and Surprising Story of the Caribbean Pirates and the Man Who Brought Them Down.* Florida: Harcourt Inc.

Wright, James Martin. (1905). *History of the Bahama Islands, with a Special Study of the Abolition of Slavery in the Colony.* Baltimore: The Friedenwald Company.

Made in the USA
Columbia, SC
12 January 2025